# Discovering the Elements

## No Simple Stories

## Analysis: Historical Cases in Chemistry

Print ISSN: 3041-0614
Online ISSN: 3041-0657

**Series Editors:** Annette Lykknes
(*Norwegian University of Science & Technology, Norway*)
Brigitte Van Tiggelen
(*Science History Institute, USA*)

---

*Published*

Vol. 2 *Discovering the Elements: No Simple Stories*
edited by Brigitte Van Tiggelen and Annette Lykknes

Vol. 1 *Green Chemistry Avant La Lettre: The Pine Institute and Resin Chemistry in Aquitaine (1900–1970)*
by Marcin Krasnodębski

Analysis: Historical Cases in Chemistry - Volume 2

# Discovering the Elements

## No Simple Stories


Editors

**Brigitte Van Tiggelen**

*Science History Institute, USA*

**Annette Lykknes**

*Norwegian University of Science and Technology, Norway*



World Scientific

NEW JERSEY • LONDON • SINGAPORE • BEIJING • SHANGHAI • HONG KONG • TAIPEI • CHENNAI • TOKYO

*Published by*

World Scientific Publishing Co. Pte. Ltd.
5 Toh Tuck Link, Singapore 596224
*USA office:* 27 Warren Street, Suite 401-402, Hackensack, NJ 07601
*UK office:* 57 Shelton Street, Covent Garden, London WC2H 9HE

Library of Congress Control Number: 2025939108

**British Library Cataloguing-in-Publication Data**
A catalogue record for this book is available from the British Library.

Cover photo
Goodyear Tire and Rubber Company. "Go Ahead on SYNTHETIC RUBBER.......the [Japanese] Might Take Malaya Some Day," circa 1942. Science History Institute, Philadelphia.
https://digital.sciencehistory.org/works/6395w729x.
Courtesy of Science History Institute.

**Analysis: Historical Cases in Chemistry — Vol. 2**
**DISCOVERING THE ELEMENTS**
**No Simple Stories**



ISBN 978-981-98-1384-1 (hardcover)
ISBN 978-981-98-1385-8 (ebook for institutions)
ISBN 978-981-98-1386-5 (ebook for individuals)

For any available supplementary material, please visit
https://www.worldscientific.com/worldscibooks/10.1142/14334#t=suppl

Typeset by Stallion Press
Email: enquiries@stallionpress.com

https://doi.org/10.1142/9789819813858_fmatter

# PREFACE

What is a scientific discovery, and what place do discoveries have in popular culture and teaching? Why examine historical discoveries, and how have they been presented as part of the story about science? How does the presentation of discoveries matter to the professional historian, chemist, student, and general public? These are the questions the contributors to this volume set out to answer through various case studies spread over several centuries, focusing on what initially seems to be the same quest: The discovery of chemical elements.

The cover photo of this volume captures some of the myths and folklore about discoveries that we wish to debunk — or at the very least challenge — in this volume: The illustration presents a proud moment when a chemist finally holds the coveted material he has been trying to produce for a long time; furthermore, this moment is dated in time in that we see a calendar in the background (dated October 1924) pointing to a defined discovery stage [Goodyear Tire and Rubber Company, n.d.]. While efforts in the laboratory clearly have been made before this occasion, the moment when the physical evidence has been produced comes as the culmination of longstanding work, and we might say that the illustration tries to separate a "before" and "after" of this great achievement — as simple and clear-cut as that. The chemist we see contemplates the special moment with a mix of triumph and joy and shares the enthusiasm (as well as the sense of achievement) with a colleague. In fact, the illustration is from a black-and-white print advertisement for Chemigum synthetic rubber and displays a research chemist (in his white laboratory collar) and an executive from the Goodyear Tire & Rubber Company (as hinted in the calendar illustration). The accompanying text promotes the research for an "all-American synthetic rubber", as Japan's expanding influence in East Asia has raised concerns about it losing access to raw materials. We note that it seems to be important for the illustrator to convey that while it ascribes the discovery to one moment, one place and one (or more) individual(s), it has to be shared and recognised as such by others. All these aspects of discovery histories, i.e., the pinpointing of events in time and place, the reduction of the process to simple and linear events, as well as the importance of sharing results to establish new knowledge in a community, are discussed in the chapters that constitute this volume.

In our experience, and as we will demonstrate in the introductory chapter of this volume, discovery histories fascinate the public and schoolchildren alike. But how can we negotiate this enthusiasm and — if we may say so — hang on to disseminate discovery tales in a simplified and heroic manner with a respect for historical scholarship that is not necessarily accessible (both in terms of actual access to work but also referring to the way the accounts are often written) to teachers and science writers? Our ambition was that the scholarly informed case stories discussed in this volume should be written in a concise and accessible fashion and be interesting for a wide range of audiences. We are also proud that each chapter comes with didactic materials (compiled at the end of the volume) for different target groups, which will hopefully make them applicable in different learning contexts.

Indeed, with the case studies in this volume and their associated didactic materials, we attempt to bridge the gap between professional historians eager to contextualise and scrutinise the depths of scientific discovery yet wanting to reach out and change public understanding of science from chemist-historians, chemists interested in the development of ideas in chemistry, and chemistry teachers "whose objective is to teach chemistry, rather than a lesson in the methodology or the sociology of science" [Jensen, 1993, p. 265] towards a common understanding of the nature of chemical discoveries focusing on elements.

This collection of case stories stems from special symposia held at the 9th Conference of the European Society for the History of Science (Bologna) in 2020 and at the 48th World Chemistry Congress organised by the International Union for Pure and Applied Chemistry (IUPAC) (Montréal) in 2021, both held virtually in pandemic times. We are grateful to the contributors for their interesting articles and for their patience with the production process. As always, we are indebted to our excellent and flexible editor, Yugarani Thanabalasingam, for her support. We also thankfully acknowledge the three reviewers (one chemist, one historian, and one chemistry educator) who provided very useful feedback on the different contributions. Brigitte Van Tiggelen wishes to specially thank Xavier Urbain for support through the finishing line.

We use a modification of the Library of Congress transliteration standard for all Russian names. Note that adherence to significant digit rules has not always been possible when presenting historical data.

Louvain-la-Neuve and Utrecht, September 2024

Brigitte Van Tiggelen Annette Lykknes

## References

Goodyear Tire and Rubber Company. (n.d.). "Go Ahead on SYNTHETIC RUBBER... the [Japanese] Might Take Malaya Some Day," circa 1942. Science History Institute. Philadelphia. https://digital.sciencehistory.org/works/6395w729x

Jensen, W. B. (1993). Challenge to public outreach. History of chemistry and the chemical community: Bridging the gap? In *Chemical Sciences in the Modern World*, ed. Mauskopf, S. H. (University of Pennsylvania Press, Philadelphia), pp. 262–276.

https://doi.org/10.1142/9789819813858_fmatter

# NOTES ON CONTRIBUTORS

**Gisela Boeck** has been associate professor at the Institute of Chemistry of the University of Rostock and a visiting professor at the Institute of Technology of Organic Chemistry, Riga Technical University. She first graduated in chemistry and specialised in quantum chemistry for her doctorate. She has taught general, inorganic and organic chemistry to medical students, as well as courses on the history of chemistry for chemistry students. She is currently chair of the "History of Chemistry" division of the German Chemical Society and has for many years represented the Society in the Working Party on the History of Chemistry (now Division of History of Chemistry, EuChemS). Her research interests concern pedagogy, didactics, and the history of chemistry. Boeck has published widely about the history of the periodic system, and her more recent publications include *Lothar Meyer: Modern Theories and Pathways to Periodicity* (Springer, 2022, with Alan Rocke).

**Carmen Giunta** is professor emeritus of chemistry at Le Moyne College in Syracuse, New York, USA. He joined the faculty there in 1990 after receiving a Ph.D. in chemical physics from Harvard University. His earliest work in the history of chemistry involved assembling the Classic Chemistry website, which he has maintained since 1996. His most recent work is *A Brief History of the Metric System: From Revolutionary France to the Constant-Based SI* (2023). He is editor of the *Bulletin for the History of Chemistry*.

**Pere Grapí** graduated in chemistry from the University of Barcelona in 1975 and completed his PhD in the history of science at the Autonomous University of Barcelona in 1996. He taught physics and chemistry at the secondary education level until his retirement in 2012. Spending some periods as an associate professor of the history of science at the Autonomous University of Barcelona, he is now an independent scholar whose main research interests are focused on the history of chemistry during the late 18$^{th}$ and early 19$^{th}$ centuries, as well as on the relation between the history of science and teaching. He is also a member of the Council of the Catalan Chemical Society. Among Grapí's more recent publications is *Inspiring Air. A History of Air-Related Science* (2019).

**Sarah Hijmans** is associate professor at the Department of History and Philosophy of Science of Universite Paris Cité (France), where she received her PhD in History and Philosophy of Science and Technology in 2023. Her dissertation, supervised by Dr. Jean-Pierre Llored, studied the concept of chemical elements in practice between 1770 and 1870 by analysing the ways in which chemists identified substances as being elementary. Hijmans has previously been a Doan Short-Term Fellow at the Science History Institute (USA), a Freer Prize Fellow at the Royal Institution of Great Britain (UK), a visiting PhD student at the University of Cambridge (UK), and a postdoctoral research and teaching associate in History of Science at the Technische Universität Berlin (Germany). Her work has been published in *Ambix*, *European Journal of Philosophy of Science* and *Foundations of Chemistry*.

**Helge Kragh** is emeritus professor at the Niels Bohr Institute, University of Copenhagen, and has previously held positions at Cornell University, USA, Aarhus University, Denmark, and the University of Oslo, Norway. His publications include articles and books on the histories of physics, chemistry, astronomy and cosmology. In 2019, he received the Roy G. Neville Prize from the Science History Institute for the biographical study *Julius Thomsen: A Life in Chemistry and Beyond*, and in the same year, he co-edited with

M. Longair the *Oxford Handbook in the History of Modern Cosmology*. His latest book, published by Oxford University press in 2024, is *The Names of Science: Terminology and Language in the History of the Natural Sciences*.

**Annette Lykknes** is a historian of chemistry and professor of chemistry education at NTNU-University of Science and Technology in Trondheim. She is chair of the Division of History of Chemistry (DHC) of the European Chemical Society (EuChemS), the History of Chemistry group of the Norwegian Chemical Society, and the Norwegian Association for History of Science. Her research interests include, among other things, the history of 19$^{th}$ and 20$^{th}$-century chemistry and university history, the history of women and couples in science, and the nature of science in chemistry textbooks. She is the book series editor (with Brigitte Van Tiggelen) of *Analysis: Historical Cases in Chemistry* (World Scientific), topic editor (history of chemistry) of *ChemTexts: The Textbook Journal of Chemistry*, and has co-edited several collective volumes in the history of science.

**Ann E. Robinson** is a historian of science who dwells in the borderlands between chemistry and physics. She received her PhD in history from the University of Massachusetts Amherst, where her dissertation examined the roles of pedagogy and research in the development of a standard form of the periodic table. Her other major project explores the controversies surrounding the discovery and naming of the superheavy elements and the role played by the International Union of Pure and Applied Chemistry in resolving the controversies. She was a fellow at the Science History Institute and the Consortium for History of Science, Technology and Medicine, both in Philadelphia. Currently, she works at the Widener Library, Harvard University.

**Charlotte A. Abney Salomon** is the Director of the Beckman Center for the History of Chemistry at the Science History Institute in Philadelphia. She received her PhD from the Program in the History of Science and Medicine at Yale University and holds an MA in

Scandinavian history from Linköping University. Her research interests centre on mineralogy, chemistry, and the natural history of minerals in 18$^{th}$-century Sweden. Previously, she taught science at the lower secondary level for six years.

**Brigitte Van Tiggelen** is Director for International Affairs and Resident Scholar at the Science History Institute, Philadelphia, USA. She chaired the Working Party on the History of Chemistry (now Division of History of Chemistry, EuChemS) between 2013 and 2021 and the Commission for the History of Chemistry and Molecular Sciences (International Union for History and Philosophy of Science and Technology) since 2017. After she graduated in physics and history, she devoted her PhD dissertation to chemistry in 18$^{th}$-century Belgium. Her research interests include topics such as couples and women in science, domestic science, the heritage of chemical and molecular sciences, philosophy of chemistry, and Belgian chemistry.

https://doi.org/10.1142/9789819813858_fmatter

# CONTENTS

https://doi.org/10.1142/9789819813858_0001

# IDENTIFYING, INTERPRETING AND DISSEMINATING DISCOVERIES OF ELEMENTS: AN INTRODUCTION TO THE VOLUME

Brigitte Van Tiggelen[1] and Annette Lykknes[2]

[1]*Mémosciences asbl, Voie du Vieux Quartier 18,*
*1348 Louvain-la-Neuve, Belgium*
*Science History Institute, 315 Chestnut Street,*
*Philadelphia, PA 19106, USA*
*bvantiggelen@sciencehistory.org*
[2]*Department of Teacher Education, NTNU-Norwegian*
*University of Science and Technology, 7491 Trondheim, Norway*
*annette.lykknes@ntnu.no*

> "Discovery! The very word sends tingles surging up your spine. It quickens your pulse. Discoveries are the moments of 'Ah, ha! I understand' and of 'Eureka! I found it!'. Everyone longs to discover something — anything!"
>
> Kendall Haven, *100 Greatest Science Discoveries of All Time* (2007)

Kendall Haven's [2007, p. xi] words from the introduction to a book about science discoveries illustrate how much discoveries fascinate. Discovery histories have a place in popular culture, where they tell the story of scientific achievements, heroes and heroines, adventure, and struggle. Titles such as *100 Greatest Science Discoveries of All Time* [Haven, 2007] and *100 Science Discoveries That Changed the World* [Salter, 2021] bear testimony to this fascination. Moreover, lists of discovery tales flourish on the Internet and include bestseller books on "morality and miracles, science and sacrifice" ([Banner, 2016] about [Patchett, 2011]), about times "when the old assumption about science, religion, and class were colliding with dangerous new ideas" ([Banner, 2016] about [Gilbert, 2014]), and about "bravery, ingenuity, and insurmountable odds" ([Banner, 2016] about [Weir, 2014]). In short, they testify to the "thrill of discovery" ([Macdougall, n.d.] about [Sobel, 2016]). Indeed, the many popular science books that include tales of discoveries bear witness to their popularity in telling how science has developed [Ball, 2002; Bryson, 2011; Emsley, 2011].

In popular storytelling, discoveries appear to be an integrant, if not an essential part, of the development of science, with milestones in an ever-rising path to understanding humans and the world. In Haven's words:

> A discovery is finding or observing something new — something unknown or unnoticed before. It is noticing what was always there but had been overlooked by all before. It is stretching out into

> untouched and uncharted regions. Discoveries open new horizons, provide new insights, and create cast fortunes. Discoveries mark the progress of human civilizations. They advance human knowledge. [Haven, 2007, p. xi]

Digging into the meaning and uses of discovery histories, we will first, in this introduction, revisit the topic of science discovery in general, its place in history, popular accounts, and teaching (Part I) before we move to the theme of our volume, the specific case of discovery of chemical elements, introduce the nine contributions on element discoveries (Part II), and discuss and draw some conclusions from these case stories (Part III).

## Part I: Why Re-Examine Scientific Discoveries?

If discoveries play such important roles in the history of science, should humans not embrace storytelling that narrates the thrill and open up to understand what it was like to "stretch out into untouched and uncharted regions", to discover new "land"? In this first part of this introduction, we will argue that discovery histories as they appear in popular accounts should not be accepted at face value but be challenged, and their mythical accounts demystified and subjected to critical examination. In the tradition so elegantly set out in the past by scholars such as Thomas Kuhn [1962], Stephen Brush [1974], Simon Schaffer [1994], and more recently by Theodore Arabatzis and Kostas Gavrolu [2016], we approach the topic by pointing to exaggerated emphasis in many discovery histories on (1) winners and losers, (2) a curated cumulative and linear thread of historical events, and (3) successors' versions of history leading up to the current state of knowledge, before we finally look at the possible pitfalls of the use of popular discovery tales in teaching.

## Reconstructing Discovery Histories, Creating Winners and Losers

The story of a scientific discovery is often told as a victory over ignorance and incomprehensibility. The military metaphor seems to apply fully here; the conceptual map of scientific thinking is redrawn — there are losers, and there are winners. Antoine-Laurent (de) Lavoisier's (1743–1794) discovery of oxygen and the ensuing concept of combustion, superseding Georg Ernst Stahl's phlogiston (1660–1734) at the end of the $18^{th}$ century, and Louis Pasteur's (1822–1895) demonstration that microorganisms were omnipresent and the cause of disease and fermentation, conquering Félix Pouchet's (1800–1872) belief in spontaneous generation, are two well-known examples [Kuhn, 1962; Bensaude-Vincent and Stengers, 1996, Chapter. 2; Collins and Pinch, 1993/1998; Waller, 2002]. On these battlefields, the names of a few are celebrated — surely Stahl and Pouchet are much less known than Lavoisier and Pasteur — even if an army of foot soldiers was enlisted in the (con)quests. In fact, in *100 Greatest Science Discoveries of All Times*, quoted above, scientific discoveries that could not be "attributed to the work of one individual or to a small group of collaborators" were excluded on purpose [Haven, 2007, p. xii], testifying to the strong inclination towards celebrating scientific achievements as the act of one genius or a small team of researchers. In the same book, Marie Curie (1867–1934) is presented as the discoverer of radioactivity, without giving any credit to her husband and collaborator Pierre Curie (1859–1906), nor to Henri Becquerel (1852–1908), with whom the Curie couple shared the Nobel Prize in Physics for the discovery of radioactivity in 1903.[1] Overall, such reconstructions of events creating winners contribute to the image that discoveries are singular moments when (some very few, if not sole) scientists observe new facts, propose new concepts, and challenge and defeat erroneous (pre)conceptions (defended by the losers) [Kuhn, 1962; Schaffer, 1986; Holmes, 2009]. These success stories, encouraged by awards

[1] See also Annette Lykknes' contribution to this volume.

such as the Nobel Prize, also reveal the nationalistic values attached to science that nations take pride in their heroes.[2]

A well-known example from chemistry, mentioned above, is the role of Lavoisier in what has become known as "the chemical revolution", and how he reconstructed the line of events and past history and made himself the author of the discovery — the winner of the battle, the hero of French chemistry [Holmes, 1985; Schaffer, 1986; Bensaude-Vincent, 1996]. The dismissal of the prevailing phlogiston theory is part of this revolutionary tale, when Lavoisier introduced a new combustion theory, replacing Stahl's combustion principle: the phlogiston, present in combustible substances and released during the combustion process. While the nature of this combustible principle was elusive for chemists for almost a century, Lavoisier discarded it and instead proposed a reversed explanation of combustion, with oxygen as the crucial reactant for that chemical process [Bensaude-Vincent and Stengers, 1996, pp. 82–91].[3] The changes this "revolution" brought with it has been so strong in traditional historiography that Frederic Lawrence Holmes, a historian of chemistry, already argued in 1989 that "[h]istorians of science have found it difficult to view 18$^{th}$-century chemistry as anything other than the stage on which the drama of the chemical revolution was performed" [Holmes, 1989, p. 3]. Indeed, winners' tales such as the story of Lavoisier defeating Stahl — Whig histories, as they are often called [Jardine, 2003], even triumphalism [Chang, 2009] — blur the complexities of the scientific endeavor and the discrepancies in interpretations at play.

## The Pitfalls of the Linear Thread

Another problematic side of traditional discovery history is the widespread use of timelines embedding these pivotal moments in which

[2] One well-known example is when the French chemist Adolphe Wurtz (1817–1884) in 1869 called chemistry a French science [Rocke, 2001]. Pasteur was also hailed as a French hero of science. See, for example, [Latour, 1993] or the recent biography of Pasteur by Michel Morange [2023].

[3] See also Pere Grapi's contribution to this volume.

linearity is the norm. Here, we refer to linearity not only in the temporal sense (as a chronology) but also epistemologically, meaning that each step in a discovery process logically follows the previous one. One might argue that discoveries redraw the conceptual map of science, as they seem to affect the progress of science in a unique way: Indeed, a new theory, a new model, a new law, a new object of enquiry, a new method, a new instrument, or a new observation appears, and novelty emerges, bringing clarity and allowing (future) scientists to build on firmer ground and explore the hitherto "unchartered regions" further. In telling these feats, one discovery leads to another, and the forged lineage between these events serves as mileposts in an articulated sequence, which appears inexorably progressive. However, research and communication processes in science rarely fit a linear model.[4] Surely, it is difficult to establish exactly how processes took place in the past, but as John Waller, a historian of medicine, warned us, not being sceptical of linear accounts of the development of science placing a single person at the centre of the progress may be seen as committing a "sin against history" [Waller, 2002, p. 284]. In the words of the philosopher Michel Serres: "Time does not always flow according to a line ... nor according to a plan but, rather, according to an extraordinarily complex mixture" ([Serres with Latour, 1995], as cited in [Tanaka, 2016, p. 176]). Indeed, instead of a linear path with gaps and associated "jumps" between them, it can be more fruitful to examine the history of science as continuous, complex and heterogeneous, sometimes with co-existing competing paradigms, and be conscious about how representations of historical developments also vary and change across time [Latour, 2007, part I, III; Allchin, 1993].

In the influential books *Laboratory Life: The Construction of Scientific Facts* [Latour and Woolgar, 1979/1986] and *Science in Action* [Latour, 1987], the sociologist and philosopher of science Bruno Latour set out to observe scientists at work from the outside, such as an anthropologist doing fieldwork in observing communities.

[4] Not to be confused with the term "linear model" used to describe the relationship between science and technology. See [Edgerton, 2004].

He aimed to understand the nature of scientific activity or science "in the making", including the unsolved controversies, to open the black boxes of science, which are often left closed in reporting scientific work. Immersed in the laboratory as the natural space of this unusual exotic tribe, Latour observed how scientists operated. However, he also experienced that when preparing accounts of the chaotic research process for publication, the scientists reformulated the process so that it appeared more linear, rational and clear-cut. Indeed, Latour established a distinction between the unfolding of the events before and after the stabilisation of the scientific fact. When applied to discoveries, this remark raises the issue of the encapsulated narratives that the sources available for historical scrutiny offer. Often, a date for a discovery is provided by the date of publication, and it is in retrospect that historical documents are investigated and interpreted [Allchin, 2013, p. 96]. This retrospective reconstruction bears the risk of painting a distorted picture of the scientific process. Moreover, the description or the claim to a discovery by their author(s) is entrenched in their own argumentation to convince their peers and, therefore, does not necessarily reflect the actual sinuous process that led to the alleged discovery, which was typically far from linear.

Another consequence of an emphasis on chronology and linear, "reported" history is that science appears to be essentially cumulative, that the new is always seen as improvements over the old, and that inventions are often seen as breakthroughs "rather than a recombination (often quite novel) of existing ideas", as stated by the historian Stefan Tanaka [2016, p. 178]. Also, when errors or detours are exposed, they are often only exhibited to prove the point that, eventually, the scientific endeavour triumphed from obscurity, bad reasoning, and unsubstantiated preconceptions, with insights of creativity supported by well-led observational and experimental investigations and logically solid conceptual elaborations. For example, in the case of Lavoisier's new theory of combustion, in retrospect, it explained why metals gained weight when burned with charcoal despite phlogiston being given off and nothing was added to the substance, and, therefore, weight loss rather than gain should have been observed. Indeed, Lavoisier used this earlier observed anomaly to present himself as the author of the discovery of what caused the weight gain [Schaffer, 1986]. Explanations existed before as well, but those who had argued that phlogiston had negative weight to get around the conceptual difficulties of phlogiston were ridiculed by supporters of the new theory [Dumas, 1837]. However, using the ability of the new theory to explain weight gain to support the supremacy of the oxygen theory is anachronistic, as weight gain was actually not observed for all substances and not even considered a problem by all contemporary phlogisticians [Kragh, 2004, pp. 215–226; Boantza and Gal, 2011; Bensaude-Vincent and Stengers, 1996, pp. 82–86]. In fact, from a modern perspective, Lavoisier's theory of combustion is not even entirely "correct": all that burn do not become an acid; moreover, combustion might take place without oxygen — and the theory involves the since long abandoned idea of "caloric" [Nicholls, 2007; Chang, 2012, pp. xviii, 8–9, 61].

## Staging Personal Involvement

Looping back to the issue of putting heroes or winners on stage in a linear thread, signalled in the two previous sections, these tales also unveil the strong personal involvement of savants and scientists and their seeking recognition. As Waller pointed out, the history of science has, up to quite recently, to a large extent, been written by scientists wanting to tell the story of their own subdiscipline and to make sure that "their part in the great drama of discovery did not get unsung" [Waller, 2002, p. 6]. Although science is usually portrayed as essentially communal, scientists who write themselves into history tend to situate themselves as creators or inventors among the community, claiming and disputing precedence or priority at the moment the discovery unfolds or when it has been acknowledged by the scientific community. Indeed, Lavoisier consistently claimed to be the designer of the new theory of chemistry after it had gained traction at the end of the 1780s and was identified as the theory of the French chemical revolution in chemistry (again reflecting that national pride was at stake as well) [Bensaude-Vincent, 1996]. Moreover, those who stage themselves as successors, compatriots or disciplinary colleagues or trainees might lead the effort of what they conceive as legitimate and due recognition for a crucial step in their science [Abir-Am, 1982]. Such thinking is reflected in commemorative practices and contemporary celebrations, sometimes with the intention that scientific commemorations should be used politically to secure the future survival of an institution [Abir-Am, 1999]. Often, the desire for a founding father is sufficiently strong to stifle their less admirable scientific work, such as when Isaac Newton (1643–1727) is depicted as the first to establish the law of universal gravitation and be a promotor of corpuscularism, without mentioning his alchemical pursuit and his theological interests [Haven, 2007, pp. 23–24; Bensaude-Vincent and Stengers, 1996, pp. 48–52; Principe, 1998], or when Charles Darwin (1809–1882) is celebrated for natural selection, while his conviction that environment can also bring structural adaptations in a Lamarckian tradition is omitted from the story [Waller, 2002, pp. 188, 288].

## Scientific Discovery in Teaching

The distorted view of discovery that follows from an emphasis on victory and defeat and cumulative events leading up to today's successful science, forged by scientists themselves, is even more problematic when such commemorative discovery histories are used to teach how science works, that is, the processes of science and how knowledge is established. This "meta-perspective" part of the curriculum is often referred to as the nature of science (NOS) and is reckoned to be a critical component of scientific literacy in school [Abd-El-Khalick and Lederman, 2023]. The problem when disseminating science is not that episodes from the history of science are not known to the general public and the students in particular, but the *kind* of histories being told. It does not help the public understanding of science or school teaching that historians and philosophers of science have since long scrutinised the instances of discovery, their parameters, and their context and analysed the factors and criteria that define this phenomenon if their work does not reach out to the broader community, especially the science teachers and their students [Lykknes, 2023]. For this reason, historian, philosopher and science educator Douglas Allchin called out not for more history but for a different kind of history, one that does not verge on myth [Allchin, 2003; 2004a].

The kind of mythical classroom stories Allchin referred to might distort the image and NOS to the extent that he called them "pseudohistory", history which contributes to spreading false ideas about science through simplified history [Allchin, 2003]. Such simplified and anecdotal, even romantic, histories are guilty of particularly three charges, which connect to the pitfalls indicated in the previous sections. Allchin stated that they (1) heroise historical actors and take their contributions out of proportions — similar to the winners and losers narratives discussed above, (2) idealise scientific processes, such as by creating linear timelines instead of acknowledging complex webs of history, and (3) sharpen actors' roles for the sake of telling a good story and, by doing so, mislead and "foster unwanted stereotypes about the nature of science" [Allchin, 2003, p. 329]. That teachers seem to like these heroic and romantic tales is part of the problem, as historian of science Stephen G. Brush discussed back in 1974. In the same way as novice scientists seem to need "authentic" life stories of heroes to "identify with and accept an idea view of the otherwise abstract and obscure workings of the social system of science" [Abir-Am, 1982, p. 282], teachers and students might think heroes in science fill an important role in teaching. Indeed, in a paper entitled "Should the history of science be rated X?" in *Science*, Brush [1974] asserted that science teachers wanted to keep their success stories and that, therefore, the (complex) history of science was unwanted because it might destroy the many "hagiographic" (heroic) tales commonly used in textbooks and classrooms. Thirty years later, Allchin wrote a follow-up article entitled "Should the sociology of science be rated X?" in *Science Education*, where he discussed how the sociology of science threatens the image of the "idealized and impersonal scientific method found in textbooks" — another contribution to depicting science in a unidimensional and simplified way [Allchin, 2004b, p. 935]. As biologist Judith Ramalay put it, "People who think science is a product rather than a messy process of inquiry can become profoundly uncomfortable when they are brought face to face with the uncertainties and arguments at the frontiers of science." ([Ramalay, 2003, p. 228], as cited in [Reiff-Cox, 2020, p. 131])

Indeed, heroic tales, anecdotes, and simplified presentations of scientific methods and processes can still be found in textbooks at different levels for the reasons we have discussed: Historical scholarship does not easily come across to teachers, and even in the cases when teachers are aware of alternative narratives, they prefer the simplified stories because they give life to stories about the development of scientific theories. As a testimony to this prevailing "grammar of textbook literature", as we may call it,[5] investigations into the presentation of the history of the periodic system in secondary school books from Spain, Norway, and Ethiopia have revealed that many texts still present the development as a linear chain of events of individual discoveries [Tesfamariam and Ejigu, 2019] and that, for instance, books fail to present the efforts to classify the chemical elements as a collective and creative endeavour [Moreno-Martínez and Lykknes, 2019]. In addition, some textbooks reinterpret the history of the periodic system to make it fit with modern scientific understanding [Kraus, 2014].[6] It is probably too optimistic to expect textbooks to demonstrate the complexity of scientific development in the tiny space often assigned to each topic by the publisher. Nevertheless, even when the opportunity arises to dedicate a lengthier segment to the historical development of a specific discovery, the pattern of heroic tales prevails.

---

[5] We owe the characterization of "grammar of literature" to Bernadette Bensaude-Vincent, who we gratefully acknowledge.

[6] While we prefer the notion of a periodic "system" over periodic "table" (because the table is just a representation of the system), please note that some authors we cite use the term "periodic table". See [Van Tiggelen *et al.*, 2019].

The excerpt below exemplifies how an American general chemistry textbook for the university level, used worldwide, chose to tell the story of how the periodic table was developed [Tro, 2016]. The whole text fills two-thirds of a page in the book, and it starts by establishing the Russian chemist Dmitrii Ivanovich Mendeleev (1834–1907) as the main scientist behind the table, although the "work of a number of chemists" contributed to the knowledge that had accumulated the 60+ chemical elements known at the time. The author continues by mentioning how Mendeleev created order by grouping elements with similar properties and listing them in order of increasing mass, going as far as predicting yet undiscovered elements:

> Mendeleev then organized all the known elements in a table consisting of rows in which mass increases from left to right. The rows were arranged so that elements with similar properties aligned in the same vertical columns. ... Mendeleev's original listing has evolved into the modern periodic table shown in [figure]. [Tro, 2016, p. 83]

The clear message is that the periodic system in the form of the table we know today "grew out of the work of Dmitrii Mendeleev", the hero of the story, who seemed to have come up with the idea of organising elements in this particular way all by himself. However, the truth is that many other chemists were working on similar classifications and are reckoned to be independent discoverers [Gordin, 2004, p. xvii; 2018; Boeck, 2019]. In fact, Mendeleev's first table did not even organise the elements from left to right as it is known today, but he did so only in his successive versions from 1871 on.[7] This work of organisation furthermore benefited from the large harvest of chemical progress on matter in general and elemental substances in particular gathered by the chemical community over the previous decades [Leal *et al.*, 2022]. Mendeleev built his systematic

---

[7] See also Gisela Boeck's contribution to this volume.

classification on other people's data. So many, in fact, that no references needed to be made since it was communal knowledge.

More profoundly, in narratives like this, where nothing is mentioned about the different processes that led to the new knowledge, including the trial and error, intuition, creativity, and imagination the work entailed [Reiff-Cox, 2020], nor about who contributed what kind of work and the amounts of time that were involved, or the kind of pieces that were needed to complete the puzzle, not much can be learnt about the nature of scientific work other than accepting that there were some geniuses who had their "ah, ha" moments and figured it all out. Such "rational reconstructions", which are merely concerned with the product of science and not the experiments, are "ultimately ahistorical in nature", Allchin [2013, p. 86] stated, and they "foster hero-fool dichotomies and associated unhealthy NOS views". As demonstrated in the excerpt above, even when some space is provided in textbooks, the habit is grounded in a pseudo-historical approach. Although case studies and classroom-adapted materials based on the history of science do exist, e.g., [Matthews, 2014; McComas, 2020; Allchin, 2013, part II; Erduran and Dagher, 2014], there is still a need for scholarly, informed stories about scientific discoveries that scrutinise the processes behind the discoveries and, in this way, help to understand the scientific practice, that is, the NOS.

## Part II: Identifying and Interpreting Discovery Histories of Chemical Elements

Nine case studies shed light on discoveries of elements in this volume. We will now present and discuss each of them. We have divided the chapters into two main groups of histories; one which particularly addresses the issue of identifying and attributing (situating) discoveries in history; the second emphasises the role of anticipation and predictions in discovery histories. We wrap up the section with Carmen Giunta's so-called "journalistic" model for describing discovery as a constructive way forward with discovery tales.

## Situating Discoveries in Time and Space

In Part I of this chapter, we justified the need to re-examine discovery histories to challenge the popular stories that often celebrate lone geniuses and draw the line of events as a line-up of sequences of developments leading to the discovery. Related to such popular history writing is the idea that discoveries often happen through "eureka" moments, brief moments in time when scientists realise that they have hatched a new idea or identified a new substance, for instance. The term stems from the archetypical legend of Archimedes (287 to ca. 212 BC), according to which he shouted "Eureka" after discovering the laws of buoyancy in his bath. A well-known example for chemists is the story about Mendeleev's discovery of the periodic law and the idea of a periodic system in 1869, which is often described as a "eureka" moment he had when playing with element cards, or an idea he conceived in his dreams [Doménech, 2016; Popova, n.d.; Gordin, 2018]. Although the textbook example we provided in Part I of this introduction does not refer to cards or dreams, the narrative is equally straightforward: "Mendeleev then [when he noticed that certain groups of elements had similar properties] organized all the known elements in a table consisting of a series of rows ..." [Tro, 2016, p. 83]. However, when discovery histories are scrutinised in context, even if there are traces of special moments of clarity, these cannot stand alone, as discoveries are always part of a scientific process, developing over time and entrenched in a certain context. Such processes are also acknowledged by scientists themselves, such as in the example of the National Aeronautics and Space Administration (NASA) intern Wolf Cukier, who, aged 17 in 2020, commented that "it's not like there's a single moment of discovery" after having discovered a previously unknown planet 1,300 light years from Earth [Helmore, 2020; Lykknes and Van Tiggelen, 2020].

Cukier's statement is even more pertinent when it comes to discoveries of elements, which is one of the reasons they have been chosen as objects of inquiry for this volume. Elements seem to form a coherent set of scientific objects, and numerous timelines give an overview of discoverers and the time they made their discovery.

One such overview, which we agree can be very useful, is the Wikipedia timeline of element discoveries [Wikipedia, n.d.]. In this timeline, there is room to give more than one name and describe which part of the discovery process with which each historical actor was involved. Another more rigid way of showing the origin of element discoveries is a "periodic table of flags", which exists in several versions on the Internet and considers the phenomenon of discovery through a specific lens, that of national pride, mentioned in Part I. Figure 1 illustrates one such table. The idea behind it is to identify one or more discoverers of each element and put the nation's flag inside the box belonging to the element they discovered to visualise or highlight each country's contributions to the periodic table. For example, the box belonging to yttrium (Z = 39) is decorated with a Finnish flag, chlorine (Z = 17) with a Swedish flag, and aluminium (Z = 13) with a Danish flag (more examples in [Lykknes and Van Tiggelen, 2020]).

Unsurprisingly, there is more to the history of the discovery of yttrium than the Finnish flag indicates. In 1794, the Finnish chemist

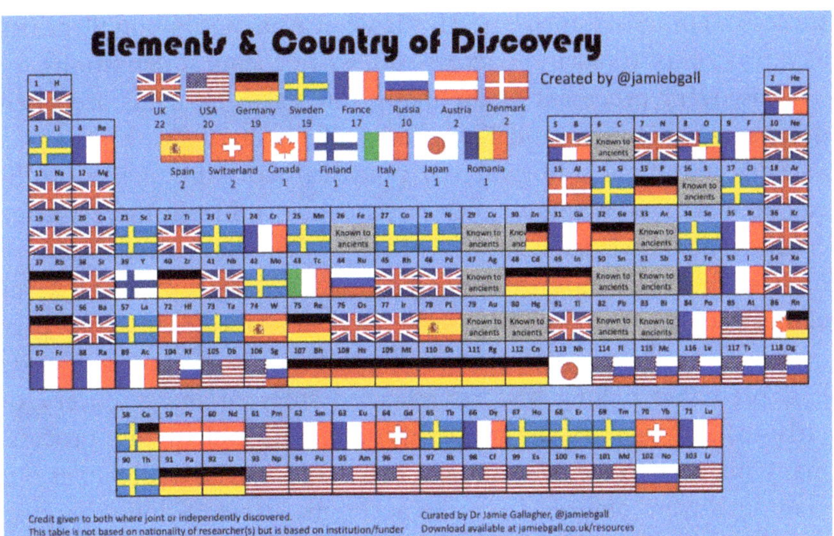


Figure 1. A periodic table of flags. Reproduced by kind permission of Jamie Gallagher.

Johan Gadolin (1750–1852) investigated a black, heavy stone from the Ytterby quarry in Sweden — whereafter he claimed that it was composed of a yet unknown earth, which he named "yttria". Gadolin's stone later became known as the mineral gadolinite, celebrating his achievements. However, yttria was later re-analysed and found to contain three distinct substances, not one.[8] Furthermore, Gadolin never isolated any yttrium metal, yet his name is still associated with the discovery of the element yttrium itself. Charlotte Abney Salomon demonstrates in her chapter in this volume, entitled "Finding Yttrium: Johan Gadolin and the Development of a Discovery", that in textbooks and mineralogical handbooks, Gadolin's substance gradually went from being a compounded substance that might have contained a metallic element to become this very metal, as yttria was gradually translated onto yttrium. The telling and retelling of the story of finding yttrium in textbooks and other publications gradually created a consensus in the chemical community, building a collective memory of Gadolin as the initiator of a collective effort to decompose an earth into its constituents parts but progressively ignoring the framework within which he and his successors practised their art. This is what Abney Salomon identifies as a process of translation across successive chemical definitions.

Discovery processes of the many rare earth elements are often quite messy [Thyssen and Binnemans, 2011; 2015], with the tours and detours of mineralogical and analytical work that make it difficult to isolate events in a coherent sequence and from there ascribe credit or acknowledgement. The credit to Gadolin for the discovery of yttrium is interesting in this respect, particularly because he never claimed credit for any discovery, and no priority dispute ever occurred. However, over time, while authors of 19$^{th}$-century texts had presented a long and complicated story of the discovery of yttrium, 20$^{th}$-century authors compressed it, and in this process, many nuances to the process were lost. Consequently, all names but Gadolin's were omitted from the accounts. Abney Salomon argues

[8] One of these substances was named terbia, and the element later extracted from this substance was named gadolinium (Z = 64).

that discoveries have become reduced to "data", with a place, date and person associated with their discovery, precisely like is seen in periodic tables of flags.

The discovery of chlorine provides another fascinating case of a changing context, both in chemical theories and practices, during which knowledge of a new substance was progressively developed. Figure 1 displays a Swedish flag in the box for chlorine, but as Pere Grapi demonstrates in his chapter "The Long and Arduous Pathway Towards the Elementary Nature of Chlorine", it took 40 years between the Swedish chemist Carl Wilhelm Scheele (1742–1786) identifying a new air and describing its characteristic properties in 1774 and the moment leading chemists agreeing to view this substance as simple, that is, indecomposable further through chemical analysis (the then established definition of an element based on Lavoisier's definition). It is also interesting that the elemental nature of chlorine was only fully acknowledged when chemists isolated an analogous substance, which was to become iodine [see also Hijmans and Llored, 2020; Hijmans, 2023, Chapter 3]. Grapi re-examines this discovery over a long period and successive modes of experimental practices and theoretical frameworks. This period saw not only a successively redefined notion of an element from the theoretical and experimental point of view but also an evolving consensus negotiated by the community. It comes as no surprise that the controversy was heated up at the moment when all the pieces of evidence of chlorine as an element were in place. Grapi shows how this was precisely the moment when priority disputes arose, and personal claims appeared vigorous. The story of chlorine also points to the fact that elemental nature does not mean existing in isolation; on the contrary, as Bernadette Bensaude-Vincent [2019] has shown, elements exist in relationship to one another. Recognizing a new chemical element demands its inclusion in a broader web of chemical interactions and analogies, which, in turn, reshapes this fragile order of the basic substances.

The discovery history of aluminium spans an even longer period, 100 years to be exact. In 1754, the Prussian chemist and pharmacist

Andreas Sigismund Marggraf (1709–1782) identified the compound alumina as a distinct chemical substance and isolated it from alum, which is sometimes reckoned to be the "discovery moment" of aluminium, while other accounts refer to the French chemist Henri Sainte-Claire Deville (1818–1881), who, in 1854, presented a new method to produce large quantities of pure aluminium to the French Academy of Sciences. In her chapter, "The Many Fathers of Aluminium: Collaboration, Credit, and the Construction of Discovery", Sarah Hijmans includes several other episodes in her web of events, among them the year 1825, when the Danish chemist Hans Christian Ørsted (1777–1851) presented a lump of aluminium at the Danish Academy of Science, which would justify the Danish flag in the periodic table in Figure 1. However, the history Hijmans presents does not stop there: Ørsted's experiments were repeated by the German chemist Friedrich Wöhler (1800–1882) shortly thereafter, and with success in 1827 and 18 years later in 1845, he produced small globules of the fusible and flexible metal. The same year that Sainte-Claire Deville invented a method to manufacture the new metal, the German chemist Robert Bunsen (1811–1899) used a similar method to produce aluminium and claimed that Sainte-Claire Deville had been inspired by him. What is, then, the answer to the question, "Who discovered aluminium"?

Hijmans shows that the answer to this question changed throughout history. At first, Wöhler was considered the first to isolate aluminium, not as a discoverer. However, when the metal was recognised as an industrially produced commodity, thanks to Sainte-Claire Deville's process, the latter reshaped the narrative through a series of priority claims in which he openly recognised Wöhler as a discoverer while excluding other contributors that had worked with him on the industrial process. To demonstrate the originality of his work, Sainte-Claire Deville detailed how his methods and results differed from Wöhler's, staging them both as fathers of aluminium. However, 30 years later, when Danish chemists in the 1920s decided to celebrate their fellow countryman in honour of the anniversary of his work on electromagnetism, they reconstructed Ørsted's experiments on aluminium and rewrote the narrative, resulting in Ørsted receiving credit in many $20^{th}$-century accounts on the history of aluminium.

The discovery histories of yttrium, chlorine and aluminium demonstrate that it is extremely difficult, if not impossible, to pinpoint a discovery to a particular time and place — or a person or country — for that matter. In the case of the discoveries of the radioactive elements polonium and radium by Marie and Pierre Curie, which Annette Lykknes discusses in her chapter entitled "A Chemistry of the Imponderable? Radium, Polonium and the Discovery of Elements in the Era of Radioactivity", the question of pinpointing discoveries deals more with which stage in the discovery process should be reckoned to constitute the discovery, rather than "who" were the discoverers. While the discovery of these elements is normally dated to 1898, the existence of polonium ($Z = 84$) was at that time only claimed on the basis of its particular radioactivity, whereas the discovery of radium ($Z = 88$) was also supported by spectroscopical evidence. No concentrated fraction containing either element had been isolated. As Lykknes shows in her reconstruction of the discovery history, the spectroscopic evidence for polonium was only provided in 1910 (visible spectra) and 1938 (X-ray spectra), respectively, and the proof typically requested to establish the existence of an element at that time, atomic weight, was obtained for radium in 1902, while it only came decades later for polonium.

Lykknes particularly emphasises the complex history of the discovery of polonium, which was the first element to be detected by radioactivity measurements, but the last to be proven by other means. Since all efforts were directed at settling the case of radium, when it finally succeeded, polonium vanished from discovery histories. In fact, it is unclear when the history of polonium's discovery ended since its atomic weight was agreed upon at the dawn of a new era of nuclear chemistry. Moreover, despite Marie Curie's stances on the new era of a "chemistry of imponderables" half a decade earlier, it was definitely by means of a more classic ponderable chemistry (analytical separation, isolation in pure metallic form, and determination of atomic weight) that the existence of the two elements was confirmed. By tracing the steps in the polonium history, Lykknes shows that even success stories such as the Curies' discoveries of the first radioactive elements were, in fact, particularly challenging and complex, the result of overlooked collaborative work, industrial connections, and continuous efforts over extensive periods.

## Predicting and Anticipating Discoveries of Elements and Isotopes

In the same way that attempts to pinpoint discoveries to particular times and places can be simplified, the way discoveries are embedded in the wider context of the development of chemistry is often undercommunicated. In fact, philosopher of science Thomas Kuhn insisted on making a difference in the historical treatment (or approach) of discoveries, depending on whether they had been predicted or not [Kuhn, 1962, p. 762]. While predicted discoveries operate within an established framework (or paradigm), unpredictable ones put that framework under tension, appearing first as anomalies or bizarreries before triggering a rearrangement of the initial framework that incorporates the anomaly. Thus, predicted discoveries appear more easily told and pinpointed. However, more often than not, discovery histories of confirmed predictions conceal doubts and discussions that occurred before discoveries could be labelled "confirmations" (if at all) [Arabatzis and Gavroglu, 2016; Kragh, 2019]. Equally complex are the processes during which elements and isotopes are named and confirmed by the scientific community. The five chapters in this volume dealing with the discoveries of the elements germanium (Z = 32), technetium (Z = 43), nobelium (Z = 102), argon (Z = 18), helium (Z = 2), and neon (Z = 10), as well as the hydrogen isotopes deuterium and tritium (Z = 1) illustrate this. We start by returning to the early versions of Mendeleev's periodic system and looking into the case of germanium.

In his 1869 version of the periodic system, like other developers of the system, Mendeleev left blank spaces for yet undiscovered elements, but unlike most others, he also described properties of these so-called *eka*-elements in some detail in his 1871 paper [Mendeleev, 1871; Gordin, 2012; 2018].[9] Traditional histories of the periodic system emphasise the significance of these predictions and how they were apparently confirmed in the 15 years following his first publication by the first three discoveries of the *eka*-elements [Brush, 1996; see also Kaji *et al.*, 2015]. However, how much did Mendeleev's predictions help chemists look for these elements, and to what extent did the subsequent discoveries confirm the predictions and the validity of the periodic system? In her chapter "Germanium — Discovered as the Predicted *Eka*-Silicon?", Gisela Boeck revisits the case of the discovery of germanium — the third *eka*-element to be discovered — through a closer examination of archival material. The question she raises is whether a new element, which was found to be different from all those known up to that point, was immediately claimed as one of the missing elements whose properties and chemical peculiarities had been predicted by Mendeleev. Indeed, the discovery of germanium by Clemens Winkler (1838–1904) in 1886 is often cited as a case of confirmation of the periodic system after gallium and scandium filled two other gaps in 1875 and 1879, respectively. Boeck argues that Mendeleev's prediction played no role in the work of discovery, identification, or isolation of germanium *per se*. Furthermore, she demonstrates that there were several discussions about placing the new metal in the periodic system and that both Winkler and Mendeleev initially did not identify germanium with *eka*-silicon — which it turned out to be, as confirmed *a posteriori*. In fact, Winkler first thought of *eka*-antimony, while Mendeleev suggested *eka*-cadmium.

Drawing on historical material, such as the correspondence between the protagonists of this debate, Boeck reveals the little-known fact that it was (Julius) Lothar Meyer (1830–1895), who had also devised a periodic system, and Victor von Richter (1841–1891),

[9] *Eka* means "one" in Sanscrit. Mendeleev used it to denote predicted elements based on their place in a group in the periodic system. *Eka*-silicon, for example, meant the first element after silicon in the same group — and translates as the first element beneath silicon in the table we know today. Likewise, "*dvi*" is used for the second, and "*tri*" for the third element in the same group.

who made the correct identification first. Mendeleev was less clear in applying the full heuristic power of his so-called periodic law. Only retrospectively, after germanium had been given the place beneath silicon in the periodic table, did Mendeleev consider Winkler along with Paul-Emile Lecoq de Boisbaudran (1838–1912) and Lars Frederik Nielsen (1840–1899; discoverers of gallium and scandium, respectively) as one of the verifiers of the periodic law. Through the finely detailed analysis, Boeck shows how later uses of the periodic system as a compass to find new elements have retrospectively altered the narrative of the discovery of germanium, with Mendeleev's effort leading the way in calling it a prediction fulfilled. However, the discussions surrounding germanium's placement inside the periodic system testify to the impulse of situating the new element in that frame, leading to a better definition of the new metal as an element with certain characteristics and the identification as *eka*-silicon, which is an element in relation with others.

In her chapter "It's not Elementary: The Successive Lives of *Eka*-Manganese aka Element 43", Brigitte Van Tiggelen discusses the discovery of another of Mendeleev's predicted *eka*-elements, element 43, or *eka*-manganese, now known as technetium. In contrast with gallium, scandium and germanium, the prediction of the Russian chemist on this unknown element was limited to a tentative atomic weight, which was indeed close to technetium's, a forecast which was, however, not included in later periodic tables by Mendeleev. Nevertheless, the discovery of this element is often presented as one of Mendeleev's successful predictions. In particular, Van Tiggelen places emphasis on the retrospective assessment of a discovery, exemplified by the case of technetium, since there were several claimed discoveries of this element that were never corroborated. The most known proposed discovery of element 43 was the one claimed by the German chemist couple Ida (1996–1978) and Walter Noddack (1893–1960), along with Otto Berg (1874-1939), who, in the same year as they announced the discovery of rhenium, Mendeleev's *dvi*-manganese, claimed to have found *eka*-manganese, which they called masurium. The discovery was never corroborated since nobody succeeded in reproducing the X-ray spectroscopical evidence. As Van Tiggelen demonstrates, attempts at rehabilitating their discovery occurred in the late 1980s and 1990s without considering the context of its discovery. While the Noddacks' search for the unknown elements was guided by analogy based on the periodic system, debates around the reinterpretation and reproduction of the Noddacks' X-ray spectroscopical evidence centred on technical aspects related to late 20$^{th}$-century nuclear physics.

Interestingly, another of the claimed discoveries of element 43, nipponium, later became associated with element 75 (rhenium), which has an atomic weight almost twice as much as that of technetium. The investigations, which were made in the late 1990s, involved both historical documents and samples; moreover, this work sparked a new and ongoing debate in the historical and philosophical scholarly communities. Again, the work of the historical actor, this time Masataka Ogawa (1865–1930), was assessed without considering the context in which it occurred, and retrospectively, Ogawa was assigned as a precursor for discovering element 75, a claim he never made. Likewise, Van Tiggelen shows that the claims for other precursors to technetium made retrospectively have been re-evaluated from the present narrator's point of view by ignoring the actors' experience and embedding their own context and perspectives into the story instead. These retrospective narratives of technetium's discovery also uncover the formidable emotional weight attached by the community to the identification of discoverers.

In her chapter entitled "Nobelium: The Questionable Element", Ann Robinson recounts the story of naming nobelium, element 102, which was less than straightforward and, therefore, provides an excellent opportunity to elaborate on the discovery process from the production of this synthetic element to the process of recognition of both its existence and its discoverers. For the first time in the history of heavy elements, nobelium's name was officialised before any announcement detailing the experiments that substantiated the discovery had been published. Indeed, the process of naming an element had shifted from that of scientists' publishing their findings to names needing authorisation from a specialised commission inside an international scientific organisation, the International Union of Pure and Applied Chemistry (IUPAC), established in 1919 [Fauque and Van Tiggelen, 2019]. In that framework, which was put in place progressively through the interwar period, the naming came as the crowning moment in the discovery process, where the approval of the community was organised through the channel of a transnational authority, which required the criteria for discovering a new element to be laid out clearly. As Robinson demonstrates, despite this seemingly well-established procedure, the Commission on Nomenclature of Inorganic Chemistry of the IUPAC in July 1957 hastily endorsed the naming of the new element announced to have been synthesised on the mere basis of telegram and correspondence. In fact, the IUPAC Commission on Atomic Weights (also referred to as the International Commission on Atomic Weights) included nobelium in its table, appearing with a name and a symbol only, despite not having an atomic weight or the indication of its most stable isotope, which was the tradition since artificially produced elements had appeared.

Problems began to surface when the first results of manufacturing nobelium could not be reproduced in their initial settings in Stockholm. When, at last, competing teams from Berkeley (USA) and Dubna (Soviet Union) indeed managed to produce element 102, each claimed the credit to have discovered it and the right to name it. While the appellation nobelium was eventually kept, with the argument that naming was not the privilege of nor acknowledgement for discovery, the heated priority dispute between the two teams revealed different criteria for assessing the discovery of a synthesised heavy element, which exceeded the expertise of the inorganic chemists. A special group, the Transfermium Working Group, was formed under the auspices of the IUPAC to adjudicate the case of nobelium and the recognition and naming of elements 101 to 109, bringing a new definition of what constituted a discovery and reinstating the connection between discovery and naming. In that sense,

Robinson argues, nobelium's story fully reveals the early steps into a new frontier of heavy elements.

In his chapter "The Heavy Hydrogen Isotopes: Anticipations and Discoveries", Helge Kragh expands the focus of discovery to isotopes, whose status borders that of elements and whose discovery stories are likewise told. He focuses on the hydrogen isotopes deuterium and tritium, which have received specific names in the IUPAC nomenclature, an unusual case for isotopes that are otherwise indicated with the corresponding mass number. His emphasis is on anticipation and surprise and the tides of predicting a possibility, which is not followed by observation or is momentarily forgotten. A focus on such cases allows one to flesh out the complex relationship between anticipation, prediction, and discovery, which, in Kuhn's view, are important aspects to be considered when approaching discoveries. Kragh states that "predictions" of discoveries extend to more than the predictions of elements based on analogies that Mendeleev and others made and include more sharp, theory-driven predictions as those known from physics and astronomy. Predictions also comprise mere speculations, a point that arises as Kragh describes the suggested existence of isotopic particles, such as deuterium and tritium. In that sense, he states, "predictions is at best an unfortunate terminology", and he considers early suggestions of deuterium not as predictions but as "loose speculations".

In the case of tritium, on the other hand, Kragh classifies Niels Bohr's (1885–1962) suggestion that there exists a kind of hydrogen with different mass as a "proper prediction" since it was based on Bohr's scientific assessment of experiments reported by the British physicist Joseph John Thomson (1856–1940). Besides, the statement had been made public, although not noted by many at the time, thus fulfilling another much-used criterion for a scientific discovery. However, Kragh does not subscribe to the idea that tritium was discovered "twice", as might sometimes be recounted because the competing research team, Kragh argues, discovered properties of the isotope and confirmed tritium's existence, which had been generally accepted by relevant scientific communities. Kragh also discusses the question of credit for discoveries of isotopes. Even if there was no doubt about the American chemist Harold C. Urey (1893–1981) being the discoverer of deuterium, it was far from self-evident that he should receive the full Nobel Prize for the discovery, a bugging issue that loops back to the many papers in this volume discussing the process of acknowledgement and fame in parallel to the process of assigning discoveries.

## Towards a New Model for Discovery History

Studying the discovery histories of argon, neon and helium, Carmen J. Giunta, in his chapter "In Defence of a Journalistic Model of Describing Discovery: The Cases of Argon, Helium and Neon" digs deeper into the question of how discovery is modelled and makes a plea for what he calls the journalistic approach to the question. This approach departs from the heroic, individual-focused narratives that have been universally criticised but draw on it in that it is based on a somewhat similar set of questions to be answered by historians. The contrasting stories of the three noble gases are investigated through the questions of "how" and "why" a discovery took place, considered alongside the more individualistic-oriented and traditional questions of "what", "when" and "where". The "how" and "why" questions allow for more depth in the investigations of past discoveries, but the problems with the three former questions, Giunta insists, do not lie with the question words themselves but with the expectations that these have simple answers. As such, this journalistic model is a useful framework for understanding the discoveries under scrutiny, and it provides the opportunity for contextualisation missing from the traditional heroic model. The "why?" question, in particular, allows much more latitude for contextualisation than the previous heroic model, elaborating on the "what" and the "how" without being Whiggish. Giunta ends his contribution by providing an answer to these questions after a careful investigation and review of all the sources at hand and contends that there is ample possibility and a useful framework for understanding these discoveries.

The three noble gases Giunta has selected to make his case are elements of a similar chemical nature, but their appearance in chemical publications and textbooks spans different periods of the development of chemistry. This allows Giunta to test his journalistic approach to the limit and gives a sense of how this more sophisticated iteration of the initial simplistic heroic model might offer a more balanced yet still summarised view on the nature of discovery in the case of chemical elements. The cases of argon, helium and neon also point to the difficulties in discovering elements that had no predicted place in the periodic system. Since they are all inert gases and, hence, rarely react with other elements, they had not been observed, and Mendeleev and others did not anticipate their existence. In fact, since argon did not seem to fit into the periodic system, there were discussions as to whether argon existed after all — or else the periodic system had to be adjusted.

## Part III: Discoveries in Chemistry — Drawing on the Case of Elements

Writing about the discovery of oxygen in 1962, James Riddick Partington (1886–1965) had a clear-cut idea of the object of his inquiry:

> The meaning of 'discovery' is well known in chemistry. When a substance is prepared for the first time by a clearly described method, and sufficient details are given which distinguish it from other substances, it is said to be 'discovered'. Glycerol was discovered by Scheele and not by Chevreul or Berthelot, ethylene by the Dutch chemists and not by Dalton. [Partington, 1962, p. 123]

A chemist-historian, Partington was also a textbook writer and an educator. As such, he was pursuing clarity of exposition. Interestingly, in his view, discovery in chemistry relates to substances and not to concepts or theories. In that respect, despite his career as a physical chemist in the highly mathematised discipline of thermodynamics, Partington seemed not to diverge from what historians and philosophers of chemistry say about the specifics of chemical sciences.

As many have argued in the last decades, chemists think with their hands and accumulate knowledge on matter and the material world by acting on it [Bensaude-Vincent and Simon, 2012; Bensaude-Vincent, 2018], and there is a definite sensuous aspect in chemistry that impacts the forms of reasoning [Chang, 2017]. For instance, early modern chemical practitioners would separate substances into their components, and the ability to reconstitute them was deemed as evidence. The successful deconstruction and reconstruction of compounds was a way of proofing for the composition, a mode of reasoning that was later refined with the interplay of the related concepts and actions of analysis and synthesis. If a new substance was produced, the novelty had to be demonstrated both by showing that it resembled no other known substances and by documenting a clear procedure that others could reproduce or inform.

One is reminded of the famous (and Promethean) declaration of the French organic chemist Marcellin Berthelot (1827–1907), "La chimie crée son objet" (Chemistry creates its object), at the same time praising the creative power of chemistry and underlining its unique mode of progress [Berthelot, 1860]. In that sense, chemistry could be seen as an exploratory science, in the way historian Freidrich Steinle [2016] approached the experimental work of the British natural philosopher Michael Faraday (1791–1867) and the French natural philosopher André-Marie Ampère (1775–1836) at the very beginning of the 19$^{th}$ century in an attempt to illustrate successful cases of non-hypothetico-deductive scientific work. Chemistry asks the question "why not?" as much as the question "why?", and is no stranger to pluralistic approaches [Chang, 2012].

With those features in mind, approaching the nature of discovery in the chemical sciences through the lens of element discoveries is less simple than it would initially appear despite the elemental character of these substances. In Kuhn's words, discoveries of phenomena are necessarily complex because they involve "recognizing both that something is and what it is" [Kuhn, 1962, p. 762]. When it comes to chemical elements, even though or because these are construed as the basic substances, Kuhn's description applies with an added twist because one does not only need to produce and identify (that is, spot or observe, isolate, identify, and procure to the rest of the community either the substance itself or the means and processes to procure it) but also prove that this new substance is of an elementary nature and had not yet been available or singled out as such [Kragh, 2019]. Since the evidence is provided over space and time, this becomes more of a long-term quest with many opportunities for debate and potential contention on who achieved what. Furthermore, the very notion of elementarity evolved through time, as is demonstrated for yttrium (Abney Salomon) and chlorine (Grapi). In other discovery histories from this volume, the shifts are more subtle but nevertheless at work. The contrasting fate of polonium and radium that are presented as being discovered in the same year as part of the same investigation (Lykknes) is one example. Challenges in establishing the discovery of new heavy and super heavy elements despite an established roadmap for accepting new synthetic elements (Robinson) are another example. In addition, the difficulties encountered by historical actors in their quest for a new element that was anachronically ignored by those who championed them as discoverers (Van Tiggelen), due to the evolving nature of what is recognised as a chemical element, further illustrate this point.

Partington's quote mentioned above features another interesting aspect: After having defined what a discovery is, he immediately moved on, without any transition, to who can be acknowledged as the discoverer. This is not the only instance of the equivalence of defining a discovery and designating the discoverer(s) in the early literature on the history of chemistry. Almost 50 years ago, the Danish chemist Edmond Rancke-Madsen published an article entitled "Discovery of element"', which provided a list of conditions that should be fulfilled should a person be credited with a discovery [Rancke-Madsen, 1976]. Clearly, the question of discovery was raised in an individualistic approach to scientific endeavour, as historians have recently emphasised [Arabatzis and Gavroglu, 2016; Kragh, 2019]. Among the criteria of what Rancke-Madsen called an "effective discovery" of an element is that the discoverer must have observed a new substance and that this substance then or later has been described as elemental. Second, the discoverer must have announced (published about) the discovery outside of their own circle. As Kragh [2019] mentioned, an important point in Rancke-Madsen's list of criteria is that a discovery cannot be a private matter known by only a small, closed circle, but the initial observation is authored and owned privately.

Thus, the traditional model of discovery based on individuals hides a tension constantly at play in the scientific endeavour: The need for consensus and common history as a background for individual achievement and recognition. In this volume, this is clearly demonstrated in Robinson's chapter, which underlines the role of the IUPAC as an enabler of diplomacy in science in times of political disconnect. In passing, Kragh also mentions the role of awards, such as the Nobel Prize, when the IUPAC does not intervene in solidifying a discovery and its author(s). Abney Salomon reports a case of no controversy in an iterative process of the construction of a consensus through decades of textbooks and technical literature on yttrium, while Hijmans points out the fact that closing the controversy on aluminium's fatherhood also served to build an identity for a community that had matured around the element.

The telling and retelling of the history of how elemental substances came to be can be seen as an expanded version of this initial reconstruction of the context of discovery. Indeed, it might appear that — in many discovery histories — once they come into the hands of their discoverers, elements become isolated from the complex network of actors or "actants" [Latour, 2005] that enabled their existence in the first place. Also, naming and predicting give existence to substances that are not yet prepared, in the way Partington presented his definition of discovery. For example, fluorine was present in all textbooks and classifications throughout the 19$^{th}$ century, appearing as early as 1816 in a list of simple substances [Caventou,

1816], yet it was only isolated in 1886 by the French chemist Henri Moissan (1852–1907). Pure ytterbium was obtained in the metallic form only in the 1950s, nearly a century after its discovery [Skelton and Thornton, 2017]. This also applies to the cases of aluminium and chlorine, not to mention yttrium and the puzzle of rare earth elements and their separation. Indeed, over time, names of the elements have been scrutinised through their etymology or their authors, with much national pride involved at times. However, naming is, to some extent, as much predicting as it is labelling, as the use of *eka*-prefixes amply shows: It brings something hinted at into existence long before it exists. Van Tiggelen unwraps this process for the case of technetium, which experienced several avatars that fuelled retrospective credit.

As discoveries were taken for granted as measures of scientific achievement or excellence and had to be allocated to individuals, the historian of science was naturally seen as a person capable of and responsible for adjudicating controversies and priority disputes. Much of the literature on discoveries has focused on this Sisyphean task. As each chapter in this volume attests, a deep dive into historical material to flesh out the discovery process might prove an inaccessible and undesirable goal for scholarship. It also appears that the very notion of discovery needs to be historicised. Indeed, it is impossible to find objective criteria for discoveries because, as the historian of science Simon Schaffer has argued, discovery is a retrospective label attributed by research communities to celebrate certain practices; therefore, it is embedded in social practice and the narrative of science [Schaffer, 1986; 1994]. It has been shown how scientific commemorations can generate new meanings about the past [Abir-Am, 1999]. This makes it difficult to find absolute criteria defining what a discovery is, once and for all, regardless of the historical context and the scientific practice. Indeed, each discovery needs to be situated at every step of constructing its narrative.

In 1938, the philosopher of science Hans Reichenbach (1895–1953) coined the distinction between the context of discovery (the processes through which novelty is produced) and the context of justification (i.e., the communication of the thinking processes and its validating among peers), insisting that epistemology only dealt with the latter [Reichenbach, 1938, pp. 5–7, 381–382]. The focus of this volume was not to discuss the abundant literature on the philosophy of discovery, but historians have built on that distinction and have continuously enriched the understanding of discovery by making use of this distinction to make sense of discovery stories [Schickore, 2022], even though, as Reichenbach rightly pointed out, epistemology, and written descriptions of scientific exposition, are in themselves reconstructions [Reichenbach, 1938, pp. 5–7, 381–383]. Several chapters in this volume strongly demonstrate that aspect of the discovery story. For example, the fine-grained analysis of Boeck on how germanium was identified and placed in the periodic system goes as close as it comes to the initial fumbling (trial and error) through both published and unpublished sources, casting doubts on the later reconstructions. Similarly, Lykknes decomposes the long and ignored path to the proof of the existence of polonium and its fundamental characterisation in terms of atomic weight and spectroscopical evidence. In addition, Kragh and Giunta, in their quest to clarify the criteria that make a discovery and the best description of it, provide a comprehensive review of hydrogen isotopes and the first three noble gases, respectively. Giunta's sophisticated contribution advocating a "journalistic model", in particular, provides a way out of the inherent tension between the need to convey condensed stories that are usable beyond the history of science scholarship and the necessity of showcasing an adequate, contextualized and substantiated view of how scientific discoveries occurred.

Beyond detailed case studies and the new perspectives they offer on the discovery stories of the respective elements, what distinguishes this volume is what it adds to the longstanding question of what a discovery is; the crucial importance of what we define as the *context of narration*. We argue that discovery is a process beyond what a linear chronology can demonstrate and that this process is constructed, deconstructed and reconstructed while being narrated, irrespective of whether the targeted community is scientists themselves, students in science, the general public, or school teachers and students. Although in the history of the science community, some of these contextual aspects have gained much attention in recent decades, they do not always come across to scientists and educators who are responsible for teaching about science and its nature to new generations. This is partly due to difficulties in aligning the goals of historians and those of scientists [Powers, 2020]. We maintain that those who have studied these processes in depth are best placed to make an effort to correct public images and narratives of science. They should take this opportunity and "move from the critical mode to the constructive mode and begin to provide some positive suggestions as to how key historical events in chemistry can be presented in the introductory textbook without offending the historical sensibilities of the professional", as chemist-historian William B. Jensen strongly stated in a contribution about the challenge to public outreach back in 1993 [Jensen, 1993, pp. 264–265]. Giunta's journalistic approach to the discovery of history provides one way of approaching such a "constructive mode". This volume, with its many varied and well-substantiated case studies and the resources for didactical use in the classroom, might provide the raw material to help address this responsibility.

## References

Abd-El-Khalick, F. and Lederman, N. G. (2023). Research on teaching, learning, and assessment of nature of science. In *Handbook of Research on Science Education Vol. III*, eds. Abell, S. K. and Lederman, N. G. (Lawrence Erlbaum Associates, Mahwah, NJ), pp. 850–898.

Abir-Am, P. G. (1982). How scientists view their heroes: Some remarks on the mechanism of myth construction. *Journal of the History of Biology*, **15**, pp. 281–315.

Abir-Am, P. G. (1999). Commemorative practices in science: Historical perspectives on the politics of collective memory. *Osiris*, **14**, pp. 1–33.

Allchin, D. (1993). Phlogiston after oxygen. *Ambix*, **39**, pp. 110–116.

Allchin, D. (2003). Scientific myth-conceptions. *Science Education*, **87**, pp. 329–351.

Allchin, D. (2004a). Pseudohistory and pseudoscience. *Science & Education*, **13**, pp. 179–195.

Allchin, D. (2004b). Should the sociology of science be rated X? *Science Education*, **88**(6), p. 934946.

Allchin, D. (2013). Teaching the Nature of Science. *Perspective & Resources* (SHiPS Education Press, Saint Paul).

Arabatzis, T. and Gavroglu, K. (2016). From discrepancy to discovery: How argon became an element. In The Philosophy of Historical Case Studies [Boston Studies in the Philosophy of Science, 319], eds. Sauer, T. and Scholl, R. (Springer, Switzerland), pp. 203–222.

Ball, P. (2002). The Elements. A Very Short Introduction (Oxford University Press, Oxford).

Banner, A. J. (2016). 8 Glorious and triumphant novels of scientific discovery (https://offtheshelf.com/2016/01/8-glorious-and-triumphant-novels-of-scientific-discovery/, last accessed 1 March 2023).

Bensaude-Vincent, B. (1996). Between history and memory: Centennial and bicentennial images of Lavoisier. *Isis*, **87**, pp. 481–499.

Bensaude-Vincent, B. (2018). Chemists without borders. *Isis*, **109**, pp. 597–607.

Bensaude-Vincent, B. and Simon, J. (2012). Chemistry. The Impure Science, 2$^{nd}$ ed. (Imperial College Press, United Kingdom).

Bensaude-Vincent, B. and Stengers, I. (1996). A History of Chemistry (Harvard University Press, Cambridge, MA, USA).

Bensaude-Vincent. B. (2019). Reconceptualizing chemical elements through the construction of the periodic system. *Centaurus*, **61**, pp. 99–310.

Berthelot, M. (1860), *Chimie organique fondée sur la synthèse, Vol.* 2 (Mallet-Bachelier, Paris), pp. 808–812.

Boantza, V. D. and Gal, O. (2011). The 'absolute existence' of phlogiston: The losing party's point of view. *British Journal for the History of Science*, **44**, pp. 317–342.

Boeck, G. (2019). Julius Lothar (von) Meyer (1830–1895) and the periodic system. *Substantia*, **3**, Suppl. 4, pp. 13–25.

Brush, S. G. (1974). Should the history of science be rated X? The way scientists behave (according to historians) might not be a good model for students. *Science*, **183**, pp. 1164–1172.

Brush, S. G. (1996). The reception of Mendeleev's periodic law in America and Britain. *Isis*, **87**, pp. 595–628.

Bryson, B. (2011). Seeing Further: The Story of Science, Discovery, and the Genius of the Royal Society (Mariner Books, Boston, MA).

Caventou, J.-B. (1816). *Nouvelle nomenclature chimique, d'après la classification adoptée par M. Thenard* (Grochard, Gabon, Paris).

Chang, H. (2009). We have never been Whiggish (about phlogiston), Centaurus, **51**, pp. 239–264.

Chang, H. (2012). Is Water $H_2O$? [Boston Studies in the Philisophy and History of Science, 293] (Springer, Switzerland).

Chang. H. (2017). What history tells us about the distinct nature of chemistry. *Ambix*, **64**, pp. 360–374.

Collins, H. and Pinch, T (1993/1998). *The Golem: What You Should Know About Science* (Cambridge University Press, Cambridge, UK).

Doménech, F. (2016). Mendeleev: Playing cards with chemistry, (https://www.bbvaopenmind.com/en/science/bioscience/mendeleev-playing-cards-with-chemistry/, last accessed 1 April 2023).

Dumas, J.-B. (1837). Leçons sur la philosophie chimique professées au Collège de France (Ébrare, Libraire).

Edgerton, D. (2004). The linear model did not exist: Reflections on the history and historiography of science and research in industry in the twentieth century, In *The Science-industry Nexus: History, Policy, Implications* (Nobel Symposium 123), eds. Grandin, K., Wormbs, N. and Widmalm, S. (Sagamore Beach, Mass.: Science History Publications), pp. 31–57.

Emsley, J. (2011). Nature's Building Blocks: An A-Z Guide to the Elements (Oxford University Press, Oxford).

Erduran, S. and Dagher, Z. R. (2014). Reconceptualizing the Nature of Science for Science Education: Scientific Knowledge, Practices and Other Family Categories [Contemporary Trends and Issues in Science Education, 43] (Springer, Switzerland).

Fauque, D. and Van Tiggelen, B., eds. (2019). *Special IUPAC 100. A* Glance At the Union History. *Special issue of Chemistry International*, **41**(3), pp. 1–62.

Gilbert, E. (2014). *The Signature of All Things* (Riverhead Books, New York).

Gordin, M. D. (2004). *A Well-Ordered Thing. Dmitrii Mendeleev and the Shadow of the Periodic Table* (Basic Books, New York).

Gordin, M. D. (2012). The textbook case of a priority dispute: D. I. Mendeleev, Lothar Meyer, and the periodic system, In Nature Engaged.

Science in Practice from the Renaissance to the Present [Palgrave Studies in Cultural and Intellectual History], eds. Biagioli, M. and Riskin, J. (Palgrave Macmillan, New York), pp. 59–82.

Gordin, M. D. (2018). Paper tools and periodic tables: Newlands and Mendeleev draw grids, Ambix, 65, pp. 30–51.

Haven, K. (2007). 100 Greatest Science Discoveries of All Times (Libraries Unlimited).

Helmore, E. (2020). 'There's something here': Teen discovers new planet while intering at NASA (https://www.theguardian.com/science/2020/jan/10/wolf-cukier-nasa-teen-intern-discovers-planet, last accessed 1 April 2023).

Hijmans, S. (2023). From Lavoisier to Mendeleev. The identification of chemical elements in practice between 1770 and 1870, PhD thesis, Université de Paris Cité.

Hijmans, S. and Llored, J. P. (2020). How to investigate the underpinnings of sciences? The case of the element chlorine. *Foundations of Chemistry*, 22(3), pp. 447–456.

Holmes, F. L. (1985). *Lavoisier and the Chemistry of Life. An Exploration of Scientific Creativity* (Madison: University of Wisconsin Press).

Holmes, F. L. (1989). *Eighteenth-Century Chemistry as an Investigative Enterprise* (University of California, Office for the History of Science and Technology, USA).

Holmes, F. L. (2009). Experimental systems, investigative pathways, and the nature of discovery. In *Models of Discovery and Creativity* [Origins: Studies in the Sources of Scientific Creativity], eds. Meheus, J. and Nickles, T. (Dordrecht: Springer), pp. 65–79.

Jardine, N. (2003). Whigs and stories: Herbert Butterfield and the historiography of science, *History of Science*, **41**, pp. 125–140.

Jensen, W. B. (1993). Challenge to public outreach. History of chemistry and the chemical community: Bridging the gap? In *Chemical Sciences in the Modern World*, ed. Mauskopf, S. H. (University of Pennsylvania Press, Philadelphia), pp. 262–276.

Kaji, M., Kragh, H. and Pallo, G., eds. (2015). *Early Responses to the Periodic System* (Oxford University Press, Oxford).

Kragh, H. (2004). *Naturerkendelse og videnskabsteori. De uorganiske videnskabers filosofi og historie* (Aarhus Universitetsforlag).

Kragh, H. (2019). Controversial elements: Priority disputes and the discovery of chemical elements. *Substantia*, **3**, Suppl. 5, pp. 79–90.

Kraus, R. (2014). Approaches to the periodic table. In *Science Teaching: The Contribution of History and Philosophy of Science*, ed. Matthews, M. R. (Routledge, New York and London), pp. 257–265.

Kuhn, T. (1962). Historical structure of scientific discovery. *Science*, 1, pp. 760–764.

Latour, B. (1987). *Science in Action. How to Follow Scientists and Engineers Through Society* (Open University Press. Milton Keynes).

Latour, B. (1993). *The Pasteurization of France* (Harvard University Press, Cambridge, MA).

Latour, B. (2005). *Reassembling the Social: An Introduction to Actor-Network Theory* (Oxford University Press, Oxford).

Latour, B. (2007). A textbook case revisited — Knowledge as a mode of existence. In *The Handbook of Science and Technology Studies*, 3rd ed., eds. Hacet, E. J., Lynch, M., Wajcman, J. and Amsterdamska, O. (MIT Press, Cambridge, MA), pp. 1–34.

Latour, B. and Woolgar, S. (1979/1986). *Laboratory Life. The Construction of Scientific Facts* (Princeton University Press).

Leal, W., Llanos, E. J., Bernal, A., Stadler, P. F., Jost, J., and Restrepo, G. (2022). The expansion of chemical space in 1826 and in the 1840s prompted the convergence to the periodic system. *Proceedings of the National Academy of Sciences*, **119**(30), e2119083119.

Lykknes, A. (2023). The history of chemistry and its audiences. Presentation for *Conversations on the History of Chemistry*, Commission on the History of Chemistry and Molecular Sciences, 1 June (https://www.chcms.org/conversations.html, last accessed 20 September 2024).

Lykknes, A. and Van Tiggelen, B. (2020). Discovery and controversy: Stories from the chemical elements, *Science in School*, **50** (https://www.scienceinschool.org/article/2020/discovery-and-controversy-stories-chemical-elements/).

Macdougall, D. (n.d.). The best books on scientific discovery and what makes scientists tick (https://shepherd.com/best-books/scientific-discovery, last accessed 1 March 2023).

Matthews, M., ed. (2014). International Handbook of Research in History, Philosophy and Science Teaching (Springer, Dordrecht).

McComas, W. F., ed. (2020). Nature of Science in Science Instruction [Science: Philosophy, History and Education] (Springer, Switzerland).

Mendeleev, D. I. (1871). Die periodische Gesetzmäßigkeit der chemischen Elemente. *Annalen der Chemie und Pharmacie*, suppl. vol. 8, pp. 133–229.

Morange, M. (2023). Pasteur (Éditions Gallimard, Paris).

Moreno-Martínez, L. and Lykknes, A. (2019). The periodic system and the nature of science: The history of the periodic system in Spanish and Norwegian secondary school textbooks. *Substantia*, **3**, Suppl. 4, pp. 61–74.

Nicholls, H. (2007). The oxygen revolution. *Chemistry World* (7 September, https://www.chemistryworld.com/news/the-oxygen-revolution/3003970.article, last accessed 20 September 2024).

Partington, J. R. (1962). The discovery of oxygen. *Journal of Chemical Education*, **39**, pp. 123–125.

Patchett, A. (2011). State of Wonder (Bloomsbury, London).

Popova, M. (n.d.). How Mendeleev invented his periodic table in a dream (https://www.themarginalian.org/2016/02/08/mendeleev-periodic-table-dream/, last accessed 1 April 2023).

Powers, J. C. (2020). The history of chemistry in chemical education. *Isis*, **111**(3), pp. 576–581.

Principe, L. (1998). The Aspiring Adept: Robert Boyle and His Alchemical Quest (Princeton University Press, New Jersey).

Ramalay, J. (2003). Building a public understanding of science: A matter of trust. In Science Literacy for the Twenty-First Century, eds. Marshall, S. P., Scheppler, J. A. and Palmisano, M. J. (Prometheus Books, New York), pp. 221–232.

Rancke-Madsen, E. (1976). The discovery of an element. *Centaurus*, **19**, pp. 299–313.

Reichenbach, H. (1938). Experience and Predication. An Analysis of the Foundations and the Structure of Knowledge (University of Chicago Press, Chicago).

Reiff-Cox, R. (2020). Exchanging the myth of a step-by-step scientific method for a more authentic description of inquiry in place. In Nature of Science in Science Instruction, ed. McComas, W. F. (Springer, Switzerland), pp. 127–139.

Rocke, A. (2001). Nationalizing Science. Adolphe Wurtz and the Battle for French Chemistry (MIT Press Cambridge, MA).

Salter, C. (2021). 100 Science Discoveries That Changed the World (Pavilion, United Kingdom).
Schaffer, S. (1986). Scientific discoveries and the end of natural philosophy. *Social Studies of Science*, **16**, pp. 387–420.
Schaffer, S. (1994). Making up Discovery (MIT Press, Cambridge, MA).
Schickore, J. (2022). Scientific Discovery. In The Stanford Encyclopedia of Philosophy, eds. Zalta, E. N. and Nodelman, U. (https://plato.stanford.edu/archives/win2022/entries/scientific-discovery/, last accessed 1 April 2023).
Serres, M. with Latour, B. (1995). Conversations on Science, Culture, and Time. Translated by Roxanne Lapidus. (University of Michigan Press, Ann Arbor).
Skelton, A. and Thornton, B. (2017). Iterations of ytterbium, Nature Chemistry, 9*(402)*. https://doi.org/10.1038/nchem.2755
Sobel, D. (2016). The Glass Universe. How the Ladies of the Harvard Observatory Took the Measure of the Stars (William Morrow, New York).
Steinle, F. (2016). Exploratory Experiments Ampère, Faraday, and the Origins of Electrodynamics (University of Pittsburg Press, Pittsburg, PA).
Tanaka, S. (2016), History without chronology. *Public Culture*, **28**(1), pp. 161–186.
Tesfamariam, G. M. and Ejigu, M. A. (2019). Are history aspects related to the periodic table considered in Ethiopian secondary school textbooks? *Substantia*, **3**, Suppl. 4, pp. 75–82.
Thyssen, P. and Binnemans, K. (2011). Accommodation of the rare earths in the periodic table: A historical analysis. In Handbook of the Physics and Chemistry of the Rare Earths 41, ed. Gschneidner, K. A. (Academic Press, Burlington), pp. 1–94.
Thyssen, P. and Binnemans, K. (2015). Mendeleev and the rare earth crisis. In Philosophy of Chemistry: Growth of a New Discipline [Boston Studies in the Philosophy of Science, 306], eds. Scerri, E. and McIntyre, L. (Springer, Switzerland), pp. 155–182.
Tro, N. (2016). Principles of Chemistry. *A Molecular Approach* (Pearson, Boston, MA).
Van Tiggelen, B., Lykknes, A. and Moreno-Martínez, L., (2019). Editorial. The Periodic System: A History of Shaping and Sharing. *Special issue of Substantia*, **3**, Suppl. 4, pp. 9–12.

Van Tiggelen, B., Lykknes, A. and Moreno-Martínez, L., eds. (2019). *The Periodic System: A History of Shaping and Sharing*. Special issue of Substantia, **3**, Suppl. 4.
Waller, J. (2002). Fabulous Science: Fact and Fiction in the History of Scientific Discovery (Oxford University Press).
Weir, A. (2014). *The Martian* (Crown Publishing).
Wikipedia. (n.d). Timeline of chemical element discoveries (https://en.wikipedia.org/wiki/Timeline_of_chemical_element_discoveries, last accessed 1 April 2023).

# Situating Discoveries in Time and Space

https://doi.org/10.1142/9789819813858_0002

# FINDING YTTRIUM: JOHAN GADOLIN AND THE DEVELOPMENT OF A DISCOVERY

Charlotte A. Abney Salomon

*Beckman Center for the History of Chemistry, Science History Institute, Philadelphia, PA 19106, USA*
*cabneysalomon@sciencehistory.org*

In 1794, Finnish chemist Johan Gadolin (1760–1852) published a paper in the *Transactions of the Royal Swedish Academy of Sciences* entitled "Examination of a black, heavy stone from Ytterby quarry in Roslagen",[1] in which he claimed that the stone in question was composed largely of a yet unknown earth. Two centuries later, reference works and textbooks commonly credit Gadolin with the discovery of the element yttrium on the basis of this paper. However, Gadolin neither found the quarry stone itself nor isolated any metal, and the substance he describes in the paper is an earth, a member of a fundamental class of material that ceased to exist in the first

[1] All translations are mine unless otherwise noted.

decades of the $19^{th}$ century. Yttrium exists in modern science as a chemical element, defined as a pure substance in which each component atom has 39 protons in its nucleus, and the story of its discovery is frequently reduced to the moment Gadolin proposed the existence of a new earth. Gadolin, however, worked within a system of chemical substances that is in many ways foreign to the one in use today, one built upon a different and shifting organisational framework and predating both the current atomic theory and the modern periodic table by decades.

Despite this, Gadolin has continuously been firmly associated with and assigned credit for the discovery. This claim has gone on to produce effects within and outside of the field of chemistry. When one of several new elements was identified within his namesake gadolinite in the late $19^{th}$ century, it was named after its source mineral, indirectly but for the first time resulting in an element named after a person — gadolinium [Lecoq de Boisbaudran, 1886]. Gadolin's achievement has become a point of Finnish national pride, celebrated with commemorative medals and a postage stamp, as shown in Figure 1.[2] More fundamentally, Gadolin is remembered and recognised over two centuries later through texts in the field of chemistry, in which his name is presented as the simple answer to the question of who discovered the element yttrium.

This essay examines the process by which Johan Gadolin became the discoverer of the first rare earth element. First, the essay follows the history of yttria, the material that Gadolin defined as a new, unidentified elemental substance and that others redefined as a combination of multiple simpler substances over the course of the $19^{th}$ century. Next, the essay traces the history of the claim that Gadolin's work represented a new discovery in the field of chemistry through its presence in both technical and general publications, from initial replication studies through up-to-date reference works and histories. Along the way, the claim to Gadolin's discovery has travelled and

[2]While Gadolin was a Swedish subject at the time of publication and wrote and worked in the Swedish language, he was born and lived in the Finnish region of Åbo, now called Turku.

Figure 1: Johan Gadolin has become a point of Finnish national pride, as illustrated by the Finnish postage stamp issued on the bicentennial of his birth, shown here in a First Day Cover. Image credit: Science History Institute, used by permission.

been transformed in a process of translation across successive chemical definitions.

## An Unknown Earth

Gadolin's 1794 paper began with: "This stone, which deserves attention for its large specific gravity [relative density] and its peculiar appearance, has been found by Herr Captain Arrhenius[3] within Ytterby Quarry ... from which a white feldspar is obtained for the porcelain works of Stockholm." For chemists in Sweden at the turn of the 19$^{th}$ century, such analysis of anomalous stones in quarries was routine; the mines were an integral component of the Swedish economy, and mineralogical chemistry was uniquely well supported, institutionalised within the government Bureau of Mines [Lindqvist, 1984; Fors, 2015]. Gadolin was not the first to examine

[3] Carl Axel Arrhenius (1757–1824), a military officer with training as a laboratory mineralogist, selected the stone in the Ytterby quarry as worthy of further investigation.

this particular stone; in the paper, he acknowledged previous chemical descriptions by Bengt Reinhold Geijer (1758–1815) in 1788 and Sven Rinman (1720–1792) in 1789. Neither had determined it to contain any unidentified substances, but Gadolin wrote that he had decided to examine it more closely because the previous trials had not provided enough information on its nature and because he happened to have been sent a sample [Gadolin, 1794].

Gadolin analysed the stone, a sample of what would come to be known as gadolinite (shown in Figure 2), in terms of its physical and chemical properties. He first detailed its appearance, including its black colour, opacity, brittleness, flavour, mass, and the character of its break, then recounted the results of a number of tests, including attempts to dissolve it in a range of acids and heating it till it melted by using a blowpipe. He declared that by weight, the mineral

Figure 2: Gadolin's 1794 paper detailed his physical and chemical analysis of a sample of a heavy black mineral, which he found to be partially composed of an unknown earth that would later be named yttria. The mineral, a modern sample of which is shown here, would eventually be dubbed gadolinite in his honour. Image credit: SBS Eclectic Images/Alamy Stock Photo. In the public domain.

consisted of "31 parts silica, 19 parts alumina, 12 parts iron oxide, and 38 parts of an unknown earth", the last of which he then isolated and chemically tested further in order to better describe it, including the results in his analysis [Gadolin, 1794, p. 146]. This description constitutes the first identification of this mineral as a distinct material with a unique set of properties and its establishment as a new naturally occurring chemical substance.

## An Elemental Earth

Gadolin was clear in his description that he was identifying three earths and a calx as components of the stone. In the system of chemistry within which Gadolin performed his investigation, calxes were known to be the elemental substance obtained after a metal (considered as a compound) had been undergoing combustion (thus a reaction with heat and air) and supposedly deprived of phlogiston. Only a few years earlier, Antoine-Laurent (de) Lavoisier (1743–1794) had proposed a new system of chemistry in which the former concept of combustion based on phlogiston was replaced with a concept based on an oxygen reaction; in the new system, the definition of a calx corresponded largely with that of an oxide. Earths, however, were a distinct and limited class of fundamental material substances in any system contemporary to Gadolin.

By the mid-18$^{th}$ century, Swedish chemists had established a definition of elements, in practice, if not yet in a codified system, as substances that could not be broken down further by any known methods [Fors, 2015, p. 100]. In the early 1780s, while Gadolin was studying at Uppsala, his mentor Torbern Bergman (1735–1784) had published multiple works in which he declared earths to be "primitive" and not further decomposable, a statement regarded to be the end of the concept of earth (singular) as an alchemical-style principle, after which the earths (multiple) were understood to be a class of basic chemical substances [Bergman, 1782; 1784]. French chemist Jean-Antoine Chaptal (1756–1832) also made clear the fundamental nature of earths in his 1790 textbook *Élémens de chimie,* declaring that all the constituent mixtures and combinations of stones in nature appeared to be made of the five primitive earths listed [Chaptal, 1796, p. 8]. In *The First Principles of Chemistry*, an English textbook published in 1790, William Nicholson (1753–1815) began his chapter on earths by defining them as substances that "are brittle, incombustible, infusible by the heat of furnaces, not soluble in several hundred times their weight of water, and destitute of metallic splendour". He identified exactly five earths: silex, clay, lime, barytes, and magnesia [Nicholson, 1790, p. 99].

The substance Gadolin described was defined as an earth according to the physical and chemical properties he detailed in his paper. However, while Gadolin made clear that he had found a substance that did not match the definition of any other known substance, he concluded with a definitive statement asserting his ambivalence about adding another earth to the list:

> All the same I dare not yet insist upon such a new discovery ... because I am of the opinion that Science would be better served if the many new earths described by Chemistry in recent years could be separated into simpler constituent parts than if their number were to be simply further increased.
>
> [Gadolin, 1794, pp. 154–55]

Despite the apparent simplicity and consistency of the era's concept and definition of an earth, the list of known earths had undergone frequent revision in the decade before Gadolin's publication, beginning almost immediately following Bergman's redefinition. In addition to the five earths that Bergman listed, the discoveries of two more were claimed before Gadolin's analysis: zircon, credited to the German chemist Martin Heinrich Klaproth (1743–1817) in 1789, and strontia, credited to Adair Crawford (1748–1795) in 1791 [Nicholson, 1808].

New earths were not immediately accepted into textbooks; authors required further justification before each change to the established list. Gadolin's new substance gained acceptance through its confirmation by repeated replication studies. In 1797, Swedish chemist Anders Gustaf Ekeberg (1767–1813) published a paper in which he confirmed Gadolin's findings and named the new earth *yttria*, after the hamlet of Ytterby in which the stone was found [Ekeberg, 1797]. Shortly thereafter, Klaproth and French chemist Nicolas Louis Vauquelin (1763–1829) each published additional confirmation and analyses of separate samples in the journal *Annales de Chimie* — Vauquelin with an introduction echoing Gadolin's, expressing his own expectation that advances in science would rectify the problem of the increasing number of earths [Klaproth, 1800; Vauquelin, 1800]. Vauquelin's paper specified that he had chosen to call the mineral *gadolinite*.[4]

As yttria made its way into chemistry texts over the next few years, the lists of earths varied widely as others were also introduced, contested or dropped. Yttria did not appear in the first edition of *A New System of Chemistry*, a textbook published by the Scottish chemist Thomas Thomson (1773–1852); Thomson included eight: lime, magnesia, barytes, strontites, alumina, silica, jargonia, and glucina [Thomson, 1800, p. 42]. Nor did the well-known French chemist Jean-Antoine-Claude Chaptal include it in the fourth edition of his *Élémens de chimie* in 1803, which maintained the same list of five (lime, magnesia, barytes, alumina, and silica) as the first edition in 1790 [Chaptal, 1803, pp. 8–14]. Most subsequent publications did include yttria, however. Despite his misgivings, Gadolin himself included it in a roster of eight earths (silica, alumina, zirconia, yttria, magnesia, barytes, lime, and strontian) in his own introductory chemistry textbook, published a few years later [Gadolin, 1798]. Within the following decade, several other new earths were also added, some of which were dropped again, as shown in Figure 3.[5]

[4] It is unclear whether he is the first to have done so; I find no uses of the name in print before this publication.

[5] For a detailed discussion of chemists' further debate on the taxonomic status of the metals resulting from the decomposition of the non-alkaline earths, see [Thyssen and Binnemans, 2011].

EARTHS.

1. Barytes.
2. Strontian.
3. Lime.
4. Magnesia.
5. Alumina.
6. Yttria.
7. Glucina.
8. Zirconia.
9. Agustina.
10. Silica.

Thomas Thomson, *A System of Chemistry*, 1802

1 Silex,
2 Alumine,
3 Lime,
4 Magnesia,
5 Barytes,
6 Strontites,
7 Zircone,
8 Glucine,
9 Yttria.

Thomas Ewell, *Plain Discourses on the Laws or Properties of Matter*, 1806.

Genus 1. Argilla
2. Magnesia
3. Calx
4. Silex
5. Strontia
6. Barytes
7. Zirconia
8. Glucina
9. Yttria
10. Agustina

James Sowerby, *British Mineralogy*, 1804

The earths, which we are to-day to examine are ten in number:

| | |
|---|---|
| SILEX, | STRONTITES, |
| ALUMINE, | YTTRIA, |
| BARYTES, | GLUCINA, |
| LIME, | ZIRCONIA, |
| MAGNESIA, | GARGONIA. |

Jane Marcet, *Conversations on Chemistry*, 1809

Figure 3: At the time that yttria was considered an elemental substance called an earth, the list of earths was variable. While some longstanding earths were universally accepted, chemists chose to include or exclude various new earths announced in chemistry journals when compiling lists for textbooks, as shown here [Thomson, 1802; Sowerby, 1804; Ewell, 1806; Marcet, 1809]. In the public domain.

## *From Yttria Earth to Yttrium Metal*

In a letter to Johan Carl Wilcke (1732–1796) on 22 May 1794, Gadolin reiterated his reluctance to declare a discovery, relating his new earth to Klaproth's zircon and making clear that though he believed himself to have discovered something new, he doubted his results would remain unchanged. He wrote, "It is not without much trepidation that I dare speak of a new earth, for they will have become all too many now if someone does not soon come across the art of further analyzing them" [Gadolin *et al.*, 1910, p. xci]. Indeed, yttria, as Gadolin described it, did not last for long. Over the next few decades, the definition of the material that Gadolin found, as well as the category of earths as a whole, changed dramatically.

At the same time that new earths were added, chemistry texts were beginning to question the entire concept.[6]

The previously straightforward category of earths quickly underwent two radical changes. In the first, chemists began to divide the earths into two groups: those that shared chemical properties with the metal salts known as alkalis and those that did not. In the 1807 *Dictionary of Chemistry and Mineralogy* entry on earths, English authors Arthur (1773–1854) and Charles Aikin (1775–1847) named nine: silex, zircon, alumine, glycine, yttria, barytes, strontian, lime, and magnesia [Aikin and Aikin, 1807]. The Aikins called for a division of the category into two; however, based on properties that only some within the group shared, proposing that if the first five remained "earths" and the remaining four became known as "alkalies", then each category would have a "genuine chemical definition" [Aikin and Aikin, 1807, p. 391]. Separate lists became widely accepted in other textbooks: alkaline earths and primitive (or true) earths, which included yttria. The 1814 American edition of *The Elements of Experimental Chemistry* by William Henry (1774–1836) listed them as such: "The alkaline earths are Barytes, Strontites, Lime, and Magnesia. The earths are Silex, Alumine, Zircon, Glucine, and Yttria" [Henry, 1814, p. 192].[7]

The second change, however, successfully challenged the existence of earths as fundamental or elemental substances. The classification was openly questioned as early as 1804, when Thomson's *System of Chemistry* included earths in the book alongside compound bodies, specifically primary compounds, rather than

[6] While the majority of the works cited in this section are English and Scottish, their authors participated actively in a cosmopolitan chemical community, and their works reflect ideas common across Europe.

[7] The author further remarks "Agustine [sic]...has been shown, by Berthollet, to be merely Phosphate of Lime".

with simple bodies such as metals [Thomson, 1804].[8] In the 1807 edition, Thomson elaborated on the decision to do so, despite the substances in question not having been physically broken down to simpler substances, remarking that their chemical similarity to many decomposable substances and dissimilarity from other categories of elemental substances made them more suited to consideration as compounds [Thomson, 1807, p. 2]. In *A Chemical Catechism*, also published in 1807, Samuel Parkes elaborated in footnotes that there was some controversy over the status of yttria specifically as an earth, as "a foreign chymist has lately announced, that when muriate of yttria is heated to redness it gives out oxygenized muriatic acid", suggesting that it may have been a metallic oxide rather than an earth, and that the English chemist Humphry Davy (1778–1829) "suspects that yttria and glucine are not distinct *primitive* earths, but modifications of other earths" [Parkes, 1807, p. 97].[9] Parkes also noted that, because the "specific gravity" [relative density] of yttria was closer to that typical of a metal than of an earth, it had "been said to be the link which connects the earths with the metals" [Parkes, 1807, p. 97].

The fundamental conceptual break came in 1808 when Davy presented a very short but hugely influential paper at the Royal Society, in which he described having used electricity to decompose the alkaline earths, producing metals, which then rapidly recomposed themselves using oxygen.[10] At the conclusion, he proposed names for the newly separated substances. Upon acceptance of his results, four fundamental substances ceased to exist, as the earth

[8] With the broad acceptance of Lavoisier's new system by this point, European chemists largely worked within its classification of materials as either simple, elemental substances or as compounds, composites of one or more simple substances.
[9] Parkes does not identify the "foreign chymist", but the claim about yttria and oxygenated muriatic acid had been published in 1805 by Swedish mineralogists Wilhelm Hisinger (1766–1852) and (Jöns) Jac.(ob) Berzelius (1779–1848) in the paper in which they announced their claim to have discovered a new metal, cerium, while attempting to isolate yttria from a mineral sample found in a mine owned by Hisinger [Hisinger and Berzelius, 1805].
[10] See also Hijmans' contribution to this volume.

barytes became known instead as an oxide of the metal barium, strontites of strontium, lime of calcium, and magnesia of what Davy ventured to call "magnium", as the name magnesium seemed already to have been used for another substance [Davy, 1894].[11]

The implications of the alkaline earth results for the primitive earths were apparent. Henry's [1814] *Elements of Experimental Chemistry*, which classified the nine earths separately from metals, as related above, discussed Davy's demonstrations as conclusive and additionally remarked in the introduction to the chapter on earths that "[t]he general tenor of the results gave great reason to conclude that alumina, silex, glucine, and zircon are, like the alkaline earths, metallic oxides; but the evidence, it must be confessed, is not altogether unequivocal" [Henry, 1814, p. 194].

In Davy's own textbook *Elements of Chemical Philosophy* [1812], he made the change explicit; there is no section for Gadolin's yttria, but instead one for its component metal element: ittrium (in his spelling) appeared as the 11$^{th}$ entry in the chapter detailing the known metals and their properties. Detailing the relationship of metals with earths, he wrote: "The most inflammable metals produce alkalies, alkaline earths, and earths in combustion. Other metals afford the substances called oxides, which are analogous to earths; and a few are converted into acids" [Davy, 1812, p. 181]. He made clear that he considered all of the substances known as earths to be oxides of metals, even those from which a metal had not been isolated, by listing silicium, aluminium, zirconium, ittrium, and glucium as "[t]he metals supposed to be contained in common earths" [Davy, 1812, p. 181]. He then described both the observed and "probable" chemical and physical properties of yttria earth.

In the preface to the Philadelphia edition of Thomson's *System of Chemistry*, the editor, Thomas Cooper (1759–1839), noted that one of the four primary "features of the modern improvements" in the new edition was "the introduction of the earths into the class of metals"

[11] Davy noted that Bergman had seemingly already claimed "magnesium" as a name for metallic manganese; manganese instead retained its name and "magnesium" indeed refers to the metal Davy derived from magnesia [Davy, 1894].

[Thomson and Cooper, 1818, pp. v–vi]. Yttrium appeared in the text as an element alongside glucinum, aluminium, and zirconium in the second of five families of metals classified as "simple combustibles". He proposed that Davy may already have isolated yttrium metal: "Davy found that when potassium is passed through red hot yttria it is converted into potash, while grey metallic particles are perceived mixed with the alkali. These it is probable are yttrium reduced to the metallic state" [Thomson and Cooper, 1818, p. 297].[12]

Not long afterwards, the process of breaking down Gadolin's discovery into simpler component parts, as he had hoped and Davy had predicted, was completed. In 1828, in a paper in the *Annales de Chimie*, German chemist Friedrich Wöhler (1800–1882) described the process he had used for isolating yttrium from yttria and glucinium from glucina, bringing the well-established idea that yttria was a metallic oxide, and no longer an irreducible substance, into material reality [Wöhler, 1828; Tansjö, 1996]. Though the resulting yttrium specifically was later contested as having been beryllium (still referred to as glucinum in 1828), the prevailing common chemical understanding had been permanently altered, and what had been Gadolin's new elemental earth was now known to be a compound substance.

## *From One Earth to Many*

In the years after earths ceased to be irreducible substances, yttria also ceased to be a single earth. In 1843, Swedish chemist Carl Gustaf Mosander (1797–1858) reanalysed yttria, finding that a specific technique of dissolution and fractional precipitation produced three distinct substances, dividing what Gadolin had called one unknown earth into a mixture of three, simultaneously adding terbia and erbia to the list of oxide earths and thus terbium and erbium to

[12] Some publications took longer to make the transition from yttria earth to the element yttrium, however. In the 1822 American edition of *Conversations on Chemistry*, a popular textbook for ladies originally published in 1805, there are nine earths, composed of "a metallic basis combined with oxygen", but in a footnote, the authors explain that they found insufficient evidence for the metallic basis of silex, alumina, yttria, glucina, and zirconia [Marcet and Comstock, 1822, p. 185].

the elements [Mosander, 1843].[13] Mosander, a former student of (Jöns) Jacob Berzelius (1779–1848), retained the original name for the most alkaline of the three resulting oxides, meaning that at this point the term "yttria" no longer referred to the material that Gadolin had described. Mosander's terbia was later further separated by other chemists, bringing ytterbia onto the list of earths in 1878, holmia, thulia, and scandia in 1879, dysprosia in 1886, and lutetia in 1907. Their component elements were now called rare earth elements, after their source, in the same fashion as the alkaline earth metals that Davy had electrified out of the alkaline earths.[14]

The new earth that Johan Gadolin wrote of in 1794 was an elemental substance, one of only a handful of component solids making up all of the world's minerals. Over time, it ceased to be one of a handful, ceased to be fundamental, and even ceased to be a single compound, as shown in Table 1. However, Gadolin retained his place in the chemistry books, as his name and his status as the discoverer, attached to yttria from the beginning, were translated onto yttrium.

Table 1: Gadolin's yttria over time. This table shows the changes in chemists' understanding of yttria, the substance that Gadolin described in 1794, over subsequent decades. While Gadolin's yttria was initially considered to be an elemental earth, it was later determined to be a metal oxide and, subsequently, the intermingled oxides of many elements.

| | |
|---|---|
| 1800 | An elemental substance, one of about 5–12 elements called earths |
| 1820 | An earth, assumed to be the oxide of a metal, as other earths have been shown to be metal oxides |
| 1830 | The oxide of a metal, yttrium, the isolation of which has been announced |
| 1850 | The intermingled oxides of three metals: yttrium, terbium and erbium |
| 1880 | The intermingled oxides of seven metals: yttrium, terbium, erbium, thulium, holmium, scandium, and ytterbium |
| 1910 | The intermingled oxides of ten metals: yttrium, terbium, erbium, thulium, holmium, scandium, ytterbium, gadolinium, dysprosium, and lutetium |

[13] Mosander published these findings in several journals simultaneously, including those cited here.
[14] The names of terbia and erbia were later exchanged.

## The Development of the Discovery

From the first appearances of yttria in print at the turn of the 19$^{th}$ century, Gadolin's name as its discoverer appeared alongside its description, first in papers discussing and confirming the finding, then in many chemistry textbooks and mineralogical handbooks. As the definition of the material changed, from earth to oxide to multiple oxides, Gadolin retained credit as the original discoverer of yttria, then yttrium, and the assertion commonly made in current literature is that Johan Gadolin discovered the element yttrium in Finland in 1794.

The discovery claim has appeared almost exclusively in three broadly defined genres of writing and has varied in generalisable ways between them. In *technical publications*, including 19$^{th}$-century scientific journals, textbooks, and chemical instruction materials, as well as 20$^{th}$-century academic papers, the discovery was generally related in a series of steps, and Gadolin frequently shared credit with others in a longer investigative process. In *non-narrative general reference works*, the discovery of yttrium has generally only been featured when a discovery claim for every element was included as a data point about each element alongside such information as its atomic mass; as such, it has been truncated and homogenised, often sufficiently to fit in a column of a largely quantitative data table. In 21$^{st}$-century histories of science written for a general audience, the claim was again related in a longer and more complicated form, but its specific content has often been further altered in the process of describing Gadolin's work within a contemporary system of chemistry.

The following passages follow the development of these claims over time, first in technical and academic publications, then in reference works and popular histories. In each genre, the work of Johan Gadolin described in his 1794 publication was translated from its original meaning and intent to fit within a new system of chemistry and a new framework of the common conception of the process of scientific discovery.

## *Technical and Academic Publications*

Following Gadolin's initial paper, the first publications that attributed a newly discovered substance to him were articles in scientific journals, beginning with the replication studies appearing in the *Annales de Chimie* in 1800, as described above. Few early authors described its finding in simple terms, with most instead recounting a process of research, usually including the contributions of others who had contributed to the discovery. Ekeberg began his paper on his confirmation of the new finding by saying, "Professor Gadolin has ... described an analysis of a black heavy stone from Ytterby quarry in Roslagen and simultaneously corrected Chemical Science through the discovery of a previously unknown and new earth" [Ekeberg 1797, pp. 156–157]. He later noted that he and Gadolin had both been inspired by the "zeal for science" of Arrhenius, who had also supplied them both with samples of the mineral for investigation. In Vauquelin's [1800] subsequent analysis, he credited Gadolin with the discovery and Ekeberg with the confirmation. Both papers, along with contributions by Klaproth, were subsequently summarised in reviews and journals in multiple languages announcing the discovery, with the credit to Gadolin intact; see [Smollett, 1802; Delamétherie, 1800; Griffiths and Griffiths, 1800; Haüy, 1801]. In one such summary, a note in an 1800 volume of the *Journal de Physique, de Chimie et d'Histoire Naturelle*, J. C. Delamétherie asked why, if the stone had been named gadolinite, the mineral had not also been named in Gadolin's honor as "gadoline" [Delamétherie, 1800].

From journals, the claim moved into chemical textbooks and academic reference works. During the $19^{th}$ century, chemistry textbook authors commonly devoted the majority of each work to detailing the physical and chemical properties of each element, substance, and substance type individually, and the discussion of the discovery of each element was often included as a fundamental portion of the element's definition. In entries on yttria, and later on yttrium, most early authors gave extensive space to the history of the substance, including to several other chemists besides Gadolin who had contributed to its study. The account in Thomson's [1804; 1818] *System of Chemistry* recounted the contributions of each of the chemists and authors involved in the discovery and isolation of the earth; no fewer than seven names (Arrhenius, Geyer, Rinman, Gadolin, Ekeberg, Vauquelin, and Klaproth) are mentioned in the 1804 edition, with Davy and Berzelius, who further analysed the material, additionally appearing by the 1818 edition. In the 1807 edition of the *Dictionary of Chemistry and Mineralogy*, which describes at length the methods and results of four subsequent analyses, the authors recounted the origin of yttria as flawed, yet accepted: "The first analysis of this mineral is that made by Professor Gadolin ... This analysis though very incorrect, deserves to be quoted, as it is the first that was made of this mineral, and as it establishes the claim of the Swedish professor to the discovery of the new earth, now called Yttria" [Aikin and Aikin, 1807, pp. 451–452]. Much later, Leopold Gmelin (1788–1853), in his *Hand-Book of Chemistry*, by now discussing the element yttrium rather than just the earth, continued the practice of naming many of the scientists whose work contributed to the discovery, stating (in contemporary translation): "Yttria was discovered in 1794 by Gadolin, and further examined by Ekeberg, Klaproth, Vauquelin, Berzelius, and Berlin. Wöhler obtained from it the metal yttrium" [Gmelin, 1843, p. 339].

Other authors, however, gave shorter versions of the story. Often, this meant that only Gadolin and Ekeberg appeared for the discovery and naming, respectively, and sometimes only Gadolin. Within his entry on yttrium, Davy began: "There is a mineral substance called Gadolinite, found at Itterby in Roslagen in Sweden, in which as peculiar earth was discovered in 1794, by Mr. Gadolin, and to which the name of *ittria* has been given" [Davy, 1812, p. 364]. In his *Manual of Chemistry*, published in London and reprinted in the United States, William Thomas Brande (1788–1866) likewise designated a section in his list of the metals to yttrium, describing the properties of yttria as observed by Berzelius via Thompson, and summarising the discovery process in a single sentence: "In 1794

Professor Gadolin discovered a new earth in a mineral from the quarry of Ytterby in Sweden, to which Ekeberg, in 1797, gave the name of Yttria" [Brande, 1819, p. 339].

For a time after 1828, when Wöhler claimed to have isolated yttrium along with beryllium and aluminium using a new technique, many textbook authors credited him with the discovery of yttrium, though some still included Gadolin as the discoverer of yttria earth, or, as it was now understood, yttrium oxide. In *Elements of Chemistry*, Benjamin Silliman (1779–1864) listed the metals aluminium, glucinum, and yttrium bracketed together, with Wöhler as the discoverer and 1828 as the date [Silliman, 1830, p. 101]. The 1831 edition of Andrew Ure's (1778–1857) *Dictionary of Chemistry and Mineralogy* included separate entries for yttria and yttrium. Yttria was described as "a new earth discovered in 1794 by Professor Gadolin, in a stone from Ytterby in Sweden" [Ure, 1831, p. 827]. Yttrium, on the other hand, "has been lately procured by M. Wöhler by a process similar to that employed for obtaining glucinum" [Ure, 1831, p. 828].

As late as the turn of the 20$^{th}$ century, some authors continued to introduce yttria with a long and complicated story. Philip E. Browning (1866–1937) gave a full paragraph to the discovery in the fourth edition of *Introduction to the Rarer Elements*, as well as a later paragraph to the work of Mosander in discussing the other rare earths, and included the clarifying statement that "[t]he term *earth* is applied to certain metallic oxides which were formerly regarded as elementary bodies" [Browning, 1919, pp. 35, 49]. Over time, however, short versions of the history of yttrium increasingly omitted all names but Gadolin in the process of conflating the earth Gadolin discovered with the metallic oxide it became, and sometimes even the metal itself, leading to the simple claim that Gadolin discovered the element yttrium. An early explicit appearance of this claim occurred in an elementary chemistry textbook published in 1851, in which he simply stated: "Yttrium was discovered in 1795, by Prof. Gadolin, in a very rare mineral, at Ytterby in Sweden, from whence its name" [Davy, 1851, p. 190].

In the early 20th century, the typical organisation of elementary chemistry textbooks changed dramatically, deemphasising the properties of individual materials, as authors centred on instruction instead on general theory and physics [see, for example, Schoch and Felsing, 1938; Sienko and Plane, 1955; Brown, 1963; Dickerson *et al.*, 1979]. As the rare earths specifically held little relevance to everyday or classroom chemistry, they were rarely discussed in textbooks, past a sentence or two addressing the lanthanides in some chapters on the periodic table, as the 20th century progressed. At the same time, historical narratives of discovery have been largely reserved only for the key figures in the development of the broader theoretical models that now take centre stage in chemistry classes, such as atomic structure, radioactivity and thermodynamics.[15]

However, in more advanced academic research and reference publications, the discovery claim for Gadolin persisted as historical background information, especially in books detailing further work on the analytical chemistry of the substance. These claims tended to appear on the first pages of books that featured them; seldom any shorter than a sentence, they were usually a part of the full story of the separation of all of the elements prefacing their technical discussion. Such claims presented the discovery as a straightforward and essential piece of information about the element while still emphasising the multi-part nature of the discovery, often showing care in ascribing contributions to multiple chemists, and thus normally maintaining that Gadolin discovered no more than yttria.

These most often described the work done in chemical terms contemporary to each publication, translating Gadolin's work into the author's chemical framework, describing Gadolin's yttria as an "oxide", and in some cases, translating multiple steps in the process across chemical systems [Spencer, 1919; Hopkins, 1923; Kleber and Love, 1963; Moeller, 1963; Callow, 1967]. For example, one author stated that Gadolin "found that the oxide occurred to the extent of 38 percent in the mineral, and resembled in its properties the oxides

[15] For an analysis of the pedagogical use of historical narrative in one such prominent textbook of the mid-20th century, see [Bensaude-Vincent, 2007].

of calcium and aluminum" [Spencer, 1919, p. 2]. Another claimed "Gadolin and Ekeberg considered that the new oxide contained a single metal" [Hopkins, 1923, p. 94]. (As discussed here, neither Gadolin nor Ekeberg considered yttria to be a metal oxide during the research to which these passages refer.)

Academic and technical authors in the 19$^{th}$ century largely emphasised the multiple steps in the process of the establishment of yttria as an earth and yttrium as an element and, in the process, distributing credit among a number of researchers. The 20$^{th}$-century authors in this genre often maintained this distributed credit with a narrative of discovery and also largely translated Gadolin's work into the terms and framework of contemporary chemistry. The primary effect of both of these actions is the demonstration of legitimacy through consensus and continuity. Authors making new or synthesised claims about the nature of chemistry, elements and yttrium, in particular, began by recapitulating its proof, establishing the legitimacy of the concept using a narrative of its history that shows the progressive acceptance in the field around the existence and definition of the material. In the later works, authors also implied the continued establishment of the substance by placing its entire history linguistically into the current system, despite the initial establishment of yttria as a substance in a category that no longer exists.

## General Reference Works and Popular Histories

The claim that Gadolin discovered the element yttrium specifically most often appeared in reference works of the late 20$^{th}$ century meant for non-specialist readers, such as dictionaries, encyclopaedias, handbooks, and websites. These works ranged from the highly technical to the broadly general and easily referenced, and they varied in design from straightforward appendix tables to infographic illustrations, as shown in Figs. 4. A fundamental feature of the claim in the genre as a whole was the presentation of the discovery of yttrium as data; in these works, all elements featured on the periodic table that were not known in antiquity had a place, date, and person associated with their discovery, expressed in similar terms for each

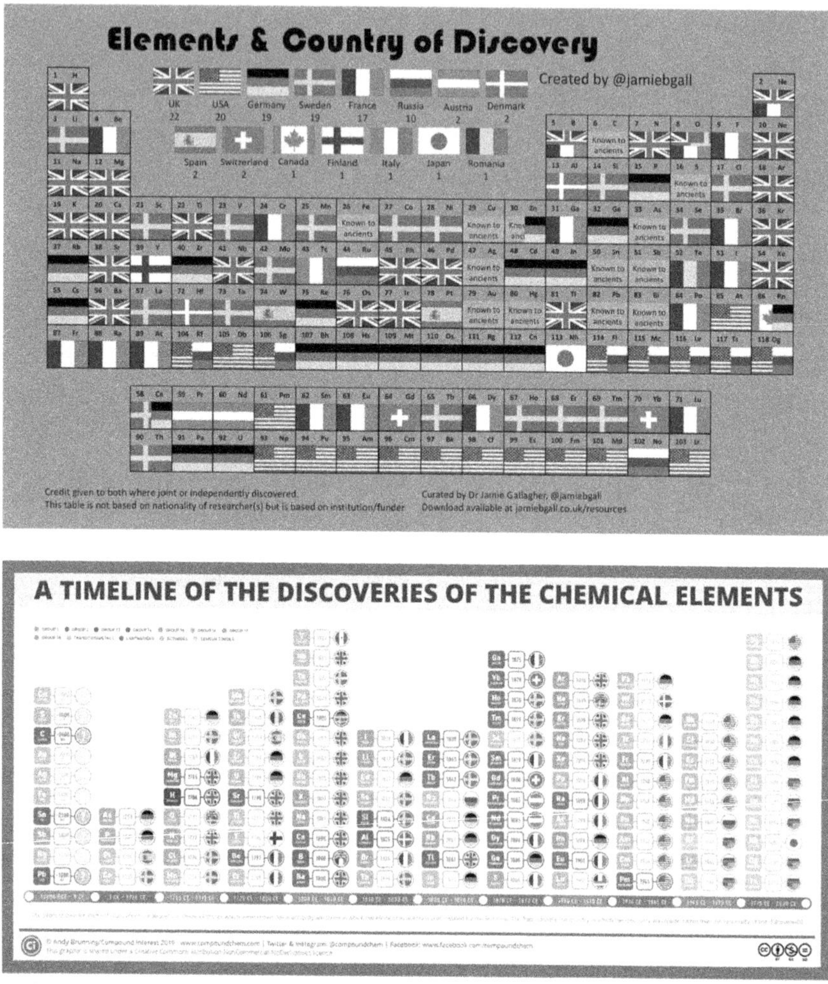


Figure 4: Discoveries of elements are also frequently seen presented in the form of infographic illustrations, as shown in these two images featured by major popular science magazines (above, *Smithsonian Magazine*, Schultz 2013; below, *American Scientist* [2019]). In these images, each element is assigned between one and three countries of origin, based on its commonly accepted discovery story, with Finland for yttrium; the lower image also includes a single year for each, with 1794 for yttrium. Image credits: Jamie Gallagher (above, used by permission[16]) and Andy Brunning (below, Creative Commons BY-NC-ND).

[16] A colour version of this table is presented in Van Tiggelen and Lykknes' introductory chapter to this volume.

one. (Interestingly, the Wikipedia page "Timeline of chemical element discoveries" has acknowledged more complexity than most, providing credit via links both for each element's observation or prediction and for its isolation [Wikipedia, 2021].)

The discovery of yttrium also appears in narrative histories of chemistry and science written for a general audience. Like early academic works, histories of chemistry and science written for a general audience in the 20th and early 21st centuries present the more complicated story for the discovery in a narrative format, but like the reference materials, they largely also recount Gadolin's work translated into modern chemistry, conflating different definitions of the substance. In some cases, the story of yttrium is presented as a heroic narrative of the importance of Gadolin's work, unknown to him at the time, and in others, a story of the progressive growth of chemical understanding.

Through many editions of *The Discovery of the Elements* by Mary Elvira Weeks (1892–1975), a highly influential text cited in a number of the publications mentioned here, the chapter on the rare earths began with sweeping and increasingly ahistorical claims for Gadolin's discovery of yttria and many other substances, including in the 1933 and 1945 editions, "The rare earths were afterward discovered in this new mineral by the famous Finnish scientist, Johan Gadolin" [Weeks, 1933, p. 230; 1948, p. 414]. The 1968 edition proclaimed, "... Lieutenant Carl Axel Arrhenius found, near the Ytterby feldspar quarry in Roslagen, an unusual black rock which he at first called ytterite, but which was later named gadolinite for the famous Finnish scientist who detected in it yttria, scandia, and all the rare earths of the yttria group" [Weeks and Leicester, 1968, p. 667].

In his work *Man and the Chemical Elements*, J. Newton Friend (1881–1966) presented the discoveries of the earths as the discoveries of all of the rare earth elements. In the text, he described the contributions of Arrhenius, Gadolin and Ekeberg in identifying yttria as a component of gadolinite, but in the accompanying chart, specified Ekeberg as the discoverer of yttrium [Friend, 1951, pp. 179–182]. Aaron J. Ihde (1909–2000), in *The Development of Modern Chemistry*, recounted the work of Arrhenius, Gadolin,

Ekeberg, and Klaproth and stated, "None of these investigators was able to separate a metal or even obtain the earth in a satisfactory state of purity" [Ihde, 1964, p. 375]. In the *Norton History of Chemistry*, William H. Brock (1936–2025) began a section on the rare earth elements by stating that they "had begun to be uncovered in 1794 when John Gadolin (1760–1852) showed that a black mineral mined at the village of Ytterby, near Stockholm, contained an unknown earth (oxide)" [Brock, 1993, p. 327]. These statements retroactively assign discovery credit for multiple $20^{th}$-century metal elements to $18^{th}$- and $19^{th}$-century chemists who understood they were working with a single nonmetallic earth.

Popular histories varied widely in their presentation of Gadolin's work, the extent of its translation, and the portion of discovery credit assigned to Gadolin. However, they shared a common framework, which is the story of the establishment of the facts of chemistry, as it is currently understood, either as a system or using the current periodic table as a structure. In these works, like those in the academic genre, Gadolin's work is not an independent achievement but serves as a beginning point for the story of the rare earth elements altogether, a starting point that is portrayed as holding within it the promise of all of the subsequent discoveries to come. Rarely is Gadolin's explicitly stated doubt about his discovery mentioned. Rather, he, along with the dozens of others described in each book, is fitted into a common modern popular conception of a scientist, defined solely in terms of the translated discovery of yttrium.

The most drastic translation of the claim involves the truncation of the claim until it consists of only three terms: Gadolin, Finland and 1794. Such works as the *McGraw Hill Encyclopedia of Science and Technology* and the reference series *The Periodic Table* stated simply that Gadolin discovered yttrium [McGraw-Hill, 1966; Saunders, 2003]. The ultimate truncation of this history into data is the appearance of the claim in data tables and infographic images, as shown in Figs. 4 and 5. (See also tables in [Collocott and Dobson, 1972; Lauerman and Alstetter, 2002; Evans, 1990].) These tables and images, while appearing most often in reference works and popular publications, also appeared in some academic works and narrative histories. In the appendix to *The Development of Modern*

*Chemistry*, yttrium is listed with the year 1794 and discoverer "Gadolin (as the impure oxide of numerous rare earths)", despite the inclusion in the text of a detailed history of yttria as an earth [Ihde, 1964, p. 747]. By the nature of their format, tables of facts about the elements show the most concise version of the claim; they very rarely name more than one researcher associated with the discovery of yttrium and nearly always name Gadolin.

The reduction of the claim in this way similarly represents the translation of Gadolin's discovery of yttria earth to yttria oxide, then to yttrium metal, and in so doing, the translation of Gadolin's work into the contemporary system of chemistry. However, in its drastic truncation, it goes on to further assert that Gadolin's identification of the earth is, in effect, the discovery of the metal, which is anachronistic. The juxtaposition of yttrium to the other elements and their discoveries presented in the same compacted format then implies that each act of identification cited is equivalent and that each of the modern elements was discovered at a single historical moment, often by a single person. This implied simplicity is belied, in fact, by the variability of the data from source to source; several give the year 1789, taken to be the year of Arrhenius's retrieval of the stone from the quarry, and some tables cite Wöhler in 1828. Both Arrhenius and Wöhler, as the first to physically find the stone and the first to claim to have isolated the metal, respectively, could be considered to have discovered yttrium under a different definition of discovery than that applied to Gadolin. In fact, many sources that cite Gadolin for yttrium cite Davy as the discoverer of each of the alkaline earth metals, specifically as the first to isolate them. There is, in fact, no consistent answer to the seemingly simple question: What constitutes the discovery of an element?

* * *

In 1794, Johan Gadolin announced that he had identified, through a series of successive laboratory tests, the presence of a substance that did not share the defining physical or chemical properties of any other currently recognised simple or compound substance but did exhibit the defining properties more broadly belonging to the class of substances known as earths. Over time, as the field of chemistry developed new classifications and subsequent chemists performed further tests, both redefined the substance in ways entirely contradictory to his initial assessment; the element yttrium, as currently defined, is in multiple ways not at all the yttria Gadolin proposed.

Johan Gadolin came to be known as the discoverer of the chemical element yttrium, not through the negotiation of contemporary scientists but through the translation of Gadolin's original work over time, as seen within written texts, both across successive systems of chemistry and definitions of matter and across genres of the publication of scientific facts. The discovery claim specifically now exists as a scientific and historical fact by virtue of its presence in these written works, therein divorced entirely from any generally applicable criteria defining a discovery, by which, if assessed, it would not likely qualify. To claim that Gadolin discovered it requires choosing the chemical analysis that pointed to an unidentified substance over the initial identification of the rock as a substance of study, the identification of the substance as the oxide of a metal (which, in fact, made the difference in the priority dispute over cerium),[17] or the subsequent redefinitions of the substance as multiple substances that include the metal yttrium.

Examination of the development over time of the discovery story suggests that Gadolin usually receives credit by virtue of being the person historically most associated with the substance. As his name and credit accompanied the accounts of his identification of yttria, and his name was also given to the mineral in which it was found (gadolinite), authors retained it through time as the earth was redefined as the oxide of an element. Even as it was divided into three component earths, Mosander's decision to keep the name yttria for one of the three kept the association of Gadolin's name as well, a retention that seems unlikely to have happened if he had not. Finally,

[17] In 1803, two simultaneous element discovery claims were made that described the same substance: Klaproth in Germany claimed to have discovered a new earth, while Hisinger and Berzelius in Sweden claimed to have discovered a new metal oxide. The priority dispute was eventually resolved in the Swedes' favour upon chemists' agreement that all earths were, in fact, oxides, and the new metal took the name the Swedes chose: cerium, after the asteroid Ceres.

each time the history of the earth was shortened to fit onto a table, nuances, complications and contributions from other chemists were lost, leaving the trio of data points: yttrium, 1794, Gadolin.

As the claim is translated into modern chemical terms, it presents a picture of chemistry in 1794 as being the same activity as chemistry in the year of publication, with the current system of chemical understanding extrapolated over 200 years into the past. A large part of the history of chemistry has been the negotiation of the taxonomy of matter. Much of the work done in the development of the chemistry in use today has been the effort to classify substances as elements, compounds or mixtures, flammable or inflammable, reactive or inert. To equate yttrium with yttria in discovery, or to give "oxide" as a direct synonym for "earth" when discussing chemistry done in the $18^{th}$ century, is to obscure a century's worth of investigation and negotiation of the nature of the earths. Additionally, the truncated version of the claim and its presentation alongside other elements suggests that discovery is a data point like any quantitative data point that is an objective measurement, necessitating a negotiation of the meaning of the word "discovery" so that it can be applied both to yttrium and to all of the other elements in the exact same way.

It would be inaccurate to say that Johan Gadolin is the discoverer of yttrium or to say that he did not discover the element and assign the credit to someone else instead; Gadolin's historical contribution to the chemical concept of yttrium is more complicated than either statement. Gadolin became in 1794 the descriptor of a new earth, as defined contemporaneously; however, over time, through translation and summarisation, he became the discoverer of the first rare earth element.

## Acknowledgements

The author wishes to thank Joanna Radin and Paola Bertucci for their guidance in the early stages of this research, as well as the participants of the Frederic L. Holmes Workshop in the Program in the History of Science and Medicine at Yale University. This essay is based on research supported by the Beckman Center for the History of Chemistry at the Science History Institute.

## References


Aikin, A. and Aikin, C. R. (1807). *A Dictionary of Chemistry and Mineralogy* (Printed for John and Arthur Arch, Cornhill; and William Phillips, London), p. 391.

American Scientist (2019). A timeline of the discoveries of chemical elements. (https://www.americanscientist.org/article/a-timeline-of-the-discoveries-of-chemical-elements, last accessed 14 April 2021).

Bensaude-Vincent, B. (2007). *College Chemistry*: How a textbook can reveal the values embedded in chemistry. *Endeavour*, 31(4), pp. 140–144.

Bergman, T. (1782). *Torberni Bergman sciagraphia regni mineralis: secundum principia proxima digesti.* (In Bibliopolio Eruditorum, Lipsiae; Dessaviae).

Bergman, T. (1784). *Physical and Chemical Essays* (Printed for J. Murray, London).

Brande, W. T. (1819). *A Manual of Chemistry* (J. Murray, London).

Brock, W. H. (1993). *The Norton History of Chemistry*, 1st ed. [Norton History of Science] (Norton, New York), p. 327.

Brown, T. L. (1963). General Chemistry. (Merrill Books, Columbus, OH, USA).

Browning, P. E. (1919). *Introduction to the Rarer Elements*, 4th ed. (J. Wiley & Sons, Inc.; Chapman & Hall, Limited, New York, London), pp. 35, 49.

Callow, R. J. (1967). *The Industrial Chemistry of the Lanthanons, Yttrium, Thorium, and Uranium* (Pergamon Press, Oxford).

Chaptal, J. -A. -C. (1796). *Elements of Chemistry: Translated from the French.* (Printed by Lang & Ustick, for selves, M. Carey [etc.], Philadelphia).

Chaptal, J. -A. -C. (1803). *Élémens de chimie*, 4th ed. (Deterville, Paris), vol. 2, p. 7.

Collocott, T. C. and Dobson, A. B. (1972). *Dictionary of Science and Technology* (Barnes & Noble, New York).

Davy, H., (1812). *Elements of Chemical Philosophy* (J. Johnson and Co., London), p. 181.

Davy, H. (1894). *The Decomposition of the Fixed Alkalies and Alkaline Earths* [Alembic club reprints] (William F. Clay, Simpkin, Marshall, Hamilton, Kent & Co. Ltd, Edinburgh, London).

Davy, J. (1851). *Chemistry and Familiar Science* (E.H. Bender, Albany, New York), p. 190.

Delamétherie, J. C. (1800). Sur une nouvelle espèce de pierre noire de Ytterby, et sur la terre particulière qui y a été trouvée. *Journal de Physique, de Chimie, et d'Histoire Naturelle,* **51**, pp. 237–239.

Diagram Group. (2006). *The Facts on File Earth Science Handbook*, revised ed. (Diagram Group, London).

Dickerson, R. E., Gray, H. B. and Haight, G. P. (1979). *Chemical Principles*, 3rd ed. (Benjamin/Cummings, Menlo Park, CA USA).

Ekeberg, A. G. (1797). Ytterligare undersökningar af den svarta stenarten från Ytterby och den däri fundna egna jord. *Kongl. Vetenskapsakademiens Nya Handlingar*, XVIII, pp. 156–164.

Evans, C. H. (1990). *Biochemistry of the Lanthanides* [Biochemistry of the Elements] (Springer, Boston, MA, USA).

Ewell, T. (1806). Plain Discourses on the Laws or Properties of Matter. (Brisban and Brannan, New York).

Fors, H. (2015). *The Limits of Matter: Chemistry, Mining, and Enlightenment* [Synthesis] (The University of Chicago Press, Chicago; London).

Friend, J. N. (1951). Man and the Chemical Elements, from Stone-Age Hearth to the Cyclotron (C. Griffin, London), 179–182.

Gadolin, J. (1794). Undersökning af en svart tung stenart ifrån Ytterby stenbrott i Roslagen. *Kongl. Vetenskapsakademiens Nya Handlingar,* XV, pp. 137–155.

Gadolin, J. (1798). *Inledning til chemien* (Frenckell, Åbo, Finland).

Gadolin, J., Hjelt, E. I. and Tigerstedt, R. (1910). *Johan Gadolin, 1760–1852: In memoriam* [Acta Societatis scientiarum fennicæ] (Druckerei der Finnischen Literaturgesellschaft, Helsingfors), p. xci.

Gmelin, L. (1848). *Hand-book of Chemistry* [Works of the Cavendish society] (Printed for the Cavendish Society, London).

Griffiths, R. and Griffiths, G. E. (1800). Chemical Annals Nos. 104–107. *The Monthly Review*, XXXIII, pp. 530–536.

Haüy, R. (1801). Notice relative to the Gadolinite. *Journal of Natural Philosophy, Chemistry, and the Arts*, 5, pp. 235–236.

Henry, W. (1814). *The Elements of Experimental Chemistry* (Thomas & Andrews, Joseph T. Buckingham, printer, Boston, MA, USA), pp. 192–194.

Hisinger, W. and Berzelius, J. J. (1805). On Cerium, a new metal found in a mineral substance of bastnas in Sweden, called tungsten. *The Philosophical Magazine*, 20, pp. 154–158.

Hopkins, B. S. (1923). *Chemistry of the Rarer Elements* (D.C. Heath and Company, Boston, New York).

Ihde, A. J. (1964). *The Development of Modern Chemistry* (Harper & Row, New York), pp. 375, 747.

Klaproth, M. (1800). Extrait d'une lettre de M. Klaproth au Cit. Vauquelin. *Annales de Chimie,* **37**, pp. 86–88.

Kleber, E. V. and Love, B. (1963). *The Technology of Scandium, Yttrium and the Rare Earth Metals* (Macmillan, New York).

Lauerman, L. and Allstetter, W. (eds.) (2002). *Science & Technology Almanac* (Greenwood Press, Westport, CT, USA).

Lecoq de Boisbaudran, P. E. (1886). Le Yα de M. De Marignac est définitivement nomme *gadolinium. Comptes Rendus Hebdomadaires des Séances de l'Académie des Sciences*, **102**, p. 899.

Lindqvist, S. (1984). *Technology on Trial: The Introduction of Steam Power Technology into Sweden, 1715–1736* (Almqvist & Wiksell, Stockholm).

Marcet, J. (1809). *Conversations on Chemistry.* (Increase Cooke & Co., New Haven).

Marcet, J. and Comstock, J. L. (1822). *Conversations on Chemistry* (Oliver D. Cooke, Hartford, CT, USA).

McGraw-Hill. (1966). *McGraw-Hill Encyclopedia of Science and Technology: An International Reference Work* (McGraw-Hill, New York).

Moeller, T. (1963). *The Chemistry of the Lanthanides* (Reinhold Pub. Corp, New York).

Mosander, C. G. (1843). On the new metals, Lanthanium and Didymium, which are associated with Cerium; and on Erbium and Terbium, new metals associated with Yttria. *Philosophical Magazine and Journal of Science,* **23**.

Nicholson, W. (1790). *The First Principles of Chemistry* (G. G. J. and J. Robinson, London), p. 99.

Nicholson, W. (1808). *A Dictionary of Practical and Theoretical Chemistry* (R. Phillips, London).

Parkes, S. (1807). *A Chymical Catechism* (J. Humphreys, Philadelphia), p. 97.

Saunders, N. (2003). *Tungsten and the Elements of Groups 3 to 7* [The Periodic Table] (Heinemann, Chicago, Ill., USA).

Schoch, E. P. and Felsing, W. A. (1938). *General Chemistry: An Introductory Course of Lessons and Exercises in Chemistry* (McGraw-Hill, New York and London).

Schultz, C. (2013). The periodic table of elemental discoveries, *Smithsonian Magazine* (https://www.smithsonianmag.com/smart-news/the-periodic-table-of-elemental-discoveries-1773011/, last accessed 14 April 2021).

Sienko, M. J. and Plane, R. A. (1955). *A Textbook of Chemistry for University Students* (Edwards Brothers, Ann Arbor, MI USA).

Silliman, B. (1830). *Elements of Chemistry* (Hezekiah Howe, New Haven, CT, USA).

Smollett, T. G. (1802). Annals of chemistry. *The Critical Review, or Annals of Literature*, XXXIV.

Sowerby, J. (1804). *British Mineralogy*, (Taylor and Company, London).

Spencer, J. F. (1919). *The Metals of the Rare Earths* (Longmans, Green, London, New York).

Tansjö, L. (1996). Carl Gustaf Mosander and his research on rare earths. In *Episodes from the History of the Rare Earth Elements* [Chemists and Chemistry] (Springer Netherlands, Dordrecht), pp. 37–54.

Thomson, T. (1800). *A New System of Chemistry*, p. 42.

Thomson, T. (1802). *A System of Chemistry*. (Bell and Bradfute [etc.], Edinburgh).

Thomson, T. (1804). *A System of Chemistry*, 2$^{nd}$ ed. (Bell and Bradfute [etc.], Edinburgh).

Thomson, T. (1807). *A System of Chemistry*, 3$^{rd}$ ed. (Bell, Bradfute & Balfour, Edinburgh), p. 2.

Thomson, T. and Cooper, T. (1818). *A System of Chemistry: In Four Volumes* (Abraham Small, Philadelphia), pp. v–vi, 193, 297.

Thyssen, P. and Binnemans, K. (2011). Accommodation of the rare earths in the periodic table: A historical analysis. In *Handbook on the Physics and Chemistry of Rare Earths*, Vol. 41, eds. Gschneidner, K. A., Bünzli, J. -C. G. and Pecharsky, V. K. (Elsevier, Amsterdam and Oxford), pp. 1–93.

Ure, A. (1831). *A Dictionary of Chemistry and Mineralogy* (Printed for T. Tegg & Son, London), pp. 827–828.

Vauquelin, N. L. (1800). Analyse d'une pierre apelée Gadolinite; Et Expose de quelques-unes des propriétés de la terre nouvelle qu'elle contient. *Annales de Chimie*, 36, pp. 143–160.

Weeks, M. E. (1933). *The Discovery of the Elements* (Mack Printing Co, Easton, PA, USA), p. 230.

Weeks, M. E. (1948). *The Discovery of the Elements* (Journal of Chemical Bulletin, Easton, PA, USA), p. 414.

Weeks, M. E. and Leicester, H. M. (1968). *The Discovery of the Elements*, 7th ed. (Journal of Chemical Education, Easton, PA, USA), p. 667.

Wikipedia. (2021). Timeline of chemical element discoveries (https://en.wikipedia.org/wiki/Timeline_of_chemical_element_discoveries, last accessed 14 April 2021).

Wöhler, F. (1828). Sur le glucinium et l'yttrium. *Annales de Chimie et de Physique*, **39**, pp. 77–84.

## Selected Bibliography

Fors, H. (2015). *The Limits of Matter: Chemistry, Mining, and Enlightenment* [Synthesis] (The University of Chicago Press, Chicago and London).

Lundgren, A. and Bensaude-Vincent, B. (eds.) (1999). *Communicating Chemistry: Textbooks and Their Audiences, 1789–1939* [European Studies in Science History and the Arts, 3] (Science History Publications, Canton, MA, USA).

Olesko, K. M. (2015). That science has been largely a solitary enterprise. In *Newton's Apple and Other Myths about Science*, eds. Numbers, R. L. and Kampourakis, K. (Harvard University Press, Cambridge, MA, USA and London, England), pp. 202–209.

https://doi.org/10.1142/9789819813858_0003

# THE LONG AND ARDUOUS PATHWAY TOWARDS THE ELEMENTARY NATURE OF CHLORINE

Pere Grapí

*Institut d'Història de la Ciència (IHC),*
*Universitat Autònoma de Barcelona, Spain*
*pgrapi@gmail.com*

Today, chlorine is a much-used chemical in the purification of drinking water, disinfectants and bleaching. Chlorine is a yellow-green gas at room temperature and is highly poisonous. For this reason, it played a notorious role in World War I, during which it was used as a chemical warfare agent, either pure or in the synthesis of phosgene, another gas weapon. Chlorine gas was even used as recently in 2007, in the Anbar Province during the Iraq War and also in the Syrian Civil War. Often credited for the discovery of this well-known halogen are the Swedish chemist Carl Wilhelm Scheele (1742–1786) and the English chemist Humphry Davy (1788–1829). Both are well-known names in the history of the discovery of chemical elements.

Joseph Priestley (1773–1804) and Antoine-Laurent (de) Lavoisier (1743–1794) are, alongside Scheele, regarded as co-discoverers of oxygen [Ihde, 1984, pp. 50–53, 66; Bensaude-Vincent and Stengers, 1993, pp. 107–110]. As early as 1772, Scheele established that common air was a mixture of two kinds of elastic fluids: "empyreal or fire-air" (oxygen), which supported combustion and calcination, and "foul or corrupted-air" (nitrogen), which did not. Scheele's findings were, however, not published in German until 1777 in his first and only monograph, *Chemische Abhandlungen von der Luft und dem Feuer* (Chemical treatise on air and fire).[1] This late publication left Scheele somewhat in the shadows of Priestley and Lavoisier. Prior to this publication, Scheele had embarked on an extensive study of the mineral pyrolusite,[2] which led to him obtaining dephlogisticated muriatic acid (later to be named "chlorine" by Davy), a previously unknown "air" (what we now know as gas). The discovery of chlorine was one case among many of what has been referred to as the hunt for "airs" that took place in Europe from the mid-18th century onward. However, the investigation into the nature of the new gas stretched beyond that period and involved many players. My aim in this contribution is to examine the complexity of this discovery process by focusing on two key aspects: (1) whether or not a particular substance was characterised as unique, and (2) whether it was understood as elemental. Because the notion of a simple substance was developed at the same time, even before that of an element, the case of chlorine provides an opportunity to understand how a conceptual framework affects the discovery process. Also of importance is the development of scientific practice and laboratory tools, which provide researchers with new ways of questioning matter.

---

[1] The monograph was translated into English in 1780 by John Reinold Forster with the title *Chemical Observations and Experiments on Air and Fire* and into French in 1781 by Baron Philippe-Frédéric de Dietrich as the *Traité chimique de l'air et du feu*.

[2] Pyrolusite is a black mineral consisting basically of manganese dioxide and is important as an ore of the manganese metal.

## Hunting Airs In the 18th Century

Natural philosophers devoted to the practice of chemistry, such as Henry Cavendish (1731–1810) and Priestley in England; medical doctors with a vocation for chemistry, such as Joseph Black (1728–1799) in Scotland; pharmacists, such as Scheele in Sweden; and lawyers-turned-chemists, such as Lavoisier in France, personified a network of diverse research sites, reinforced by the growing number of scientific publications, which encouraged the exchange of information on their chemical practices, and notably on the airs that they were investigating [Bensaude-Vincent and Stengers, 1993, pp. 106–107].

The endeavour to collect airs began at the University of Edinburgh, where Black conducted experiments with limestone (calcium carbonate), quicklime (calcium oxide) and slaked lime (calcium hydroxide) during the 1750s. These experiments led him to conclude that, apart from atmospheric air, a different kind of air existed that was contained in *magnesia alba* (basic magnesium carbonate) and in other calcareous materials.[3] He called it "fixed air" (carbon dioxide) because it was fixed in *magnesia alba* from which it could be released. Black's publication is often associated with the beginnings of gas chemistry — a special field of investigation called pneumatic chemistry — and as the bedrock in the history of gases [Black, 1756; Donovan, 1975, pp. 201–221].

[3] An abridged description of some of Black's experiments is outlined below. The calcination of both limestone and magnesia alba produced the release of fixed air [in modern terms: $CaCO_3$ (s) → $CaO$ (s) + $CO_2$ (g); $MgCO_3$ (s) → $MgO$ (s) + $CO_2$ (g)]. Slaked lime (caustic) in the presence of salt of tartar (potassium carbonate) recovered the fixed air and became limestone again [$Ca(OH)_2$ (aq) + $K_2CO_3$ (aq) → $CaCO_3$ (s) + 2KOH (aq)], which also meant the loss of its caustic character. To recover the original *magnesia alba* from its calcined (non-caustic) product, this product was first dissolved by the spirit of vitriol (sulfuric acid) [$MgO$ (s) + $H_2SO_4$ (aq) → $MgSO_4$ (aq) + $H_2O$ (l)], after which the salt of tartar made it possible to recover magnesia alba with all its original properties [$MgSO_4$ (aq) + $K_2CO_3$ (aq) → $MgCO_3$ (s) + $K_2SO_4$ (aq)]. From these experiments, Black was able first to conclude that causticity was an intrinsic property of substances liable to be calcined, and, second, that fixed air was a kind of air different from common air.

A distinguished forerunner in collecting gases was the clergyman and natural philosopher Stephen Hales (1677–1761). From 1708 until his death, he served as a minister in a parish in Teddington (Middlesex). Hales held that all gases were not more than airs and that their most interesting property was elasticity (i.e., compressibility). However, air could contain alien particles from acids or salts that attracted the air particles and altered its elasticity. In 1727, in order to remove these impurities and to estimate the amount of "air" given off by heating various solid substances, Hales devised an apparatus to "wash" airs by passing them through water to intercept impurities before collecting them in an inverted receiver (Figure 1).

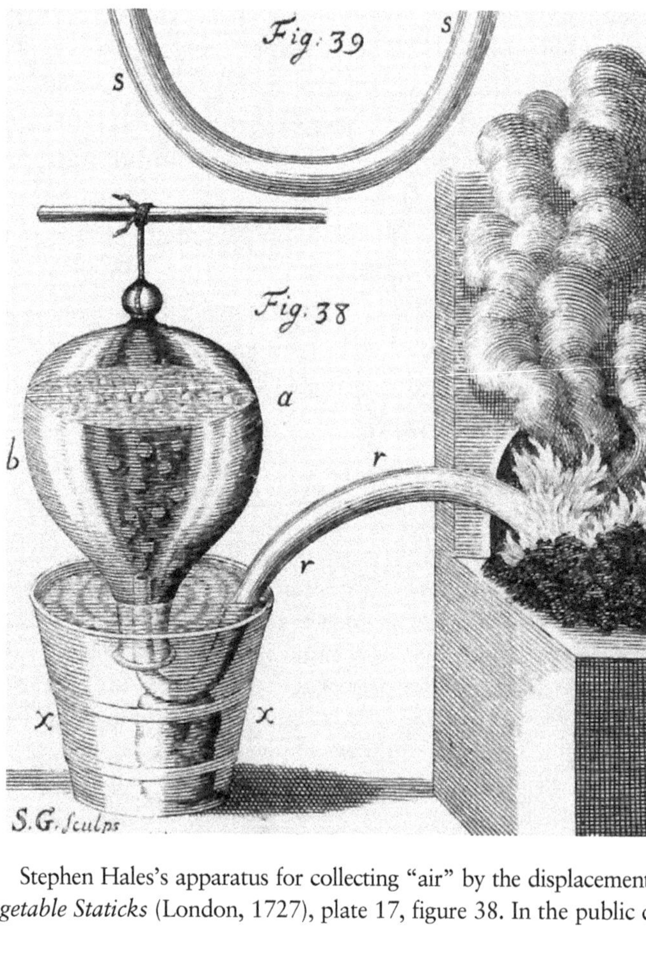


Figure 1: Stephen Hales's apparatus for collecting "air" by the displacement of water. From *Vegetable Staticks* (London, 1727), plate 17, figure 38. In the public domain.

Hales' apparatus would later serve as a prototype for Priestley's pneumatic trough [Crosland, 2000, pp. 83–85].

Although, at first, the airs continued to be interpreted as mere variants of atmospheric air, pneumatic chemists increasingly noticed that the different kinds of air differed from common air. This reinterpretation process turned out to be long and confusing despite the growing number of undeniable experimental results [Brock, 1992, pp. 81–84; Blumenthal and Ladyman, 2017]. Clearly, observational or experimental facts in science mean nothing by themselves. In the end, what make them meaningful are the mindsets of the experimenters. In the mid-18th century, chemists not only relied on corpuscular models and chemical affinities to understand the reactions observed in the laboratory, but they also made use of the theory of phlogiston for interpreting the composition of substances.[4] The latter, however, would be of little help for the chemical differentiation of airs.

Phlogiston was one of the subtle, invisible, imponderable and property-bearing fluids that arose in the context of 18th-century natural philosophy.[5] This postulated concept became the keystone for an entire theory that provided rational and coherent explanations for various phenomena but especially for the calcination of metals and combustion, or fermentation. The theory emerged in the context of the metallurgical practices in the German states and provided an explanation of the calcination of metals and the combustion

[4] During the 18th century, chemical affinities became a coherent system that provided an explanation for chemical change, founded on the assumption that they were constant and elective. Chemical affinity was interpreted as a Newtonian, short-range attraction that tended to link particles of different substances to generate new products. The result of a chemical change was predetermined by the elective order established between two reacting substances in what was known as the tables of affinity; it reflected the belief of the savants of the 18th century in a concept of nature governed by strict and invariable laws. Throughout the 18th century, these tables offered the possibility of seeing the elective order of the affinities of certain substances and, consequently, were believed to provide visible evidence of such a conception of chemical change [Sadoun-Goupil, 1991; Duncan, 1996].

[5] Phlogiston appeared on the list of these subtle fluids along with magnetism, electricity, heat, and gravity (often described as an "ether").

of coal by the release of the phlogiston contained in these materials [Brock, 1992, pp. 78–84; Bensaude-Vincent and Stengers, 1993, pp. 83–85].[6]

In 1766, Cavendish, the prototype of the wealthy and self-taught natural philosopher devoted exclusively to experimental work, published his first memoir on an inflammable air (hydrogen), obtained by the action of the dilute spirit of vitriol (sulfuric acid) on a metal like zinc, iron or tin, which he identified with phlogiston itself or with water united to phlogiston [Cavendish, 1766, pp. 144–145; Brock, 1992, p. 104; Bensaude-Vincent and Stengers, 1993, pp. 106, 109–110].[7] The point here is that, as Black had done earlier with fixed air, Cavendish recognised inflammable air as a substance different from common air. Cavendish was skilful in designing experimental devices to transfer a gas from one recipient — a bottle or a bladder — to another. Particular manual dexterity was required when pouring gas upwards from a bottle with a wide mouth into another with a narrow mouth through a funnel underwater (Figure 2).

## Perfecting The Tools of Pneumatic Chemistry

Black and Cavendish's findings encouraged Priestley to conduct a systematic study of many airs. While Priestley was living in England, he enjoyed a patronage that provided him with an annual income, which enabled him to acquire instrumentation specific to pneumatic chemistry [Schofield, 2004, pp. 3–5].[8] Those devices demanded a

[6] For instance, tin was thought to be a mixture of tin calx and phlogiston. Upon heating, the phlogiston escaped and the ash that remained was tin calx. A source of phlogiston was coal, and since it burnt away and left no ash, it was thought to be almost pure phlogiston.

[7] These three metals react with dilute sulfuric acid, releasing hydrogen. For instance:

$$\text{Zn (s)} + H_2SO_4\text{ (aq)} \rightarrow ZnSO_4\text{ (aq)} + H_2\text{ (g)}$$

[8] Although Priestley used mainly kitchen utensils such as an earthenware trough, pots and dishes, he had to purchase glassware such as vials with expensive ground-glass stoppers [Crosland, 2000, pp. 91–93].

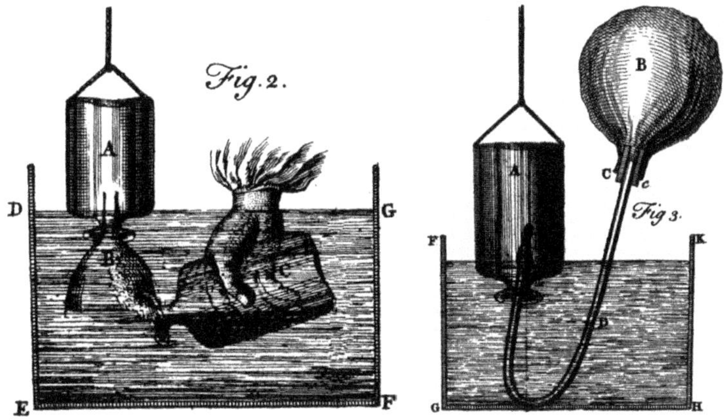


Figure 2: Cavendish's device for transferring gases. From *Philosophical Transactions* (1766), vol. 56, plate 7, p. 141. In the public domain.

certain skill to operate, which he learned to master, and very soon, he was able to observe and describe a copious collection of airs (more than 20 different gases).[9] These airs were usually collected by means of a pneumatic trough (Figure 3).

Priestley himself acknowledged that his pneumatic trough was, in fact, nothing more than the experimental devices of Hales and Cavendish diversified and made a little simpler. The apparatus consisted of a wooden oval-shaped recipient filled with water in which, at one end, there was a wooden shelf under the water. To store the several kinds of air, Priestley used cylindrical glass jars, which, once removed from the trough, he placed in pots or tea dishes of different sizes. The procedure consisted of plunging the dish into the water and sliding the jar into the dish, thereby enabling him to take both

[9]Priestley's publications on that topic appeared between 1772 and 1780: They include papers published in *Philosophical Transactions* in 1774, 1775 and 1776; *Experiments and Observations on Different Kinds of Air*, published in three volumes (1774–1775, 1776 and 1777) and its continuation *Experiments and Observations Relating to Various Branches of Natural Philosophy* (1779, 1781 and 1786), as well as the last three-volume abridged edition of the preceding six-volume series *Experiments and Observations on Different Kinds of Air and other Branches of Natural Philosophy* (1790).

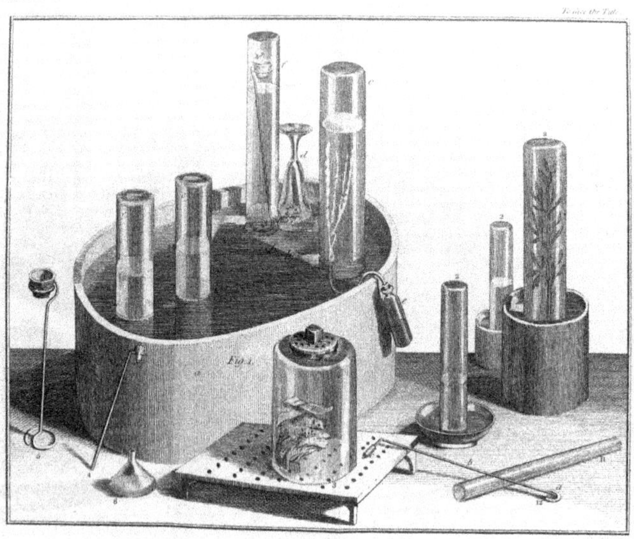

Figure 3: Joseph Priestley's pneumatic trough. From *Experiments and Observations of Different Kinds of Air* (London, 1775), plate facing the title page. In the public domain.

out together. Tea dishes were used above all for transferring a jar from one place to another. When a small pot had to be supported at a considerable height within a jar, a wire stand was used.

In order to generate and collect air by the dissolution of metals by acids or any other process, materials were put into a small flask prepared with a bent narrow glass tube inserted through a perforated cork and sealed airtight. The end of this glass tube was introduced beneath the mouth of a cylindrical jar, which was placed on the shelf and into which the released air was going to be collected. To collect those kinds of air, which were soluble in water, Priestley benefitted from Cavendish's innovation of using mercury instead of water. In this way, even airs that are particularly water-soluble, such as what we now call ammonia and hydrogen chloride, could be isolated with the aid of the pneumatic trough.

Finally, to test whether a mouse could live in any kind of air, he found that the hollow part of a tall beer glass filled with the air was very suitable for this purpose. The mouse was passed through the water and into the glass holding it by the back of the neck. According to Priestley, if the air was good, the mouse would soon feel at ease, having suffered nothing by its passage through the water. A plant could also be placed in a vessel containing any kind of air. In order to ensure that the mice were comfortable, they were placed inside a glass jar open at the top and bottom, stood on a tin plate, and covered with another plate of the same kind. Both the lower and upper plates were perforated, which allowed for ventilation [Priestley, 1775, pp. 6–22].

In his publication, Priestley proceeded like a natural philosopher, observing and collecting facts that he faithfully reported and which allowed him to demonstrate the individuality of these different airs by exploring their characteristic properties. He proceeded further by resorting to the phlogiston theory, of which he was a staunch supporter, in order to explain the nature of these different substances. This intellectual framework prompted him to develop an interpretation of different kinds of air in terms of different proportions of phlogiston in atmospheric air. Thus, phlogisticated air, such as nitrogen or carbon dioxide, was completely saturated with phlogiston and was not appropriate for breathing.[10] On the other hand, dephlogisticated air was divested of more than its normal allocation of phlogiston and was, therefore, more wholesome. Common air consisted basically of a mixture of phlogisticated (nitrogen) and dephlogisticated air (oxygen). An inflammable air, such as hydrogen, methane or carbon monoxide, contained a high proportion of phlogiston.

Glass flasks were mostly used to collect airs. They proved to be highly advantageous for carrying out experiments such as combustion since, for this purpose, animal bladders were not so suitable and were not so widely used. Nevertheless, Scheele used a bladder for collecting his "fire-air" and, among others, an as-yet unknown yellow-green air.

---

[10]With regard to animal respiration, the prevailing belief was that expired air conveyed the phlogiston released from the lungs during breathing and evacuated it from the body.

## Scheele's Isolation of a New Air

Scheele (Figure 4) worked at various apothecaries' shops in a country economically dependent on its mineralogical resources, which accounts for the great emphasis laid on chemistry by the Swedish state. He began his apprenticeship at a pharmacy in Gothenburg, and shortly after obtaining his pharmacy journeyman's certificate in 1765, he moved to a pharmacy in Malmö. At that time, he was already a skilled laboratory chemist who had read all the chemistry books he could lay his hands on [Fors, 2003, p. 171]. In 1768, he settled in Stockholm to take up employment at a pharmacy, and two years later, found a new job as a laboratory assistant at the Upland's Arms pharmacy in Uppsala [Fors, 2008, pp. 34–39]. It was in this city that Scheele connected with Torbern Olof Bergman (1735–1784), who held the chair of chemistry at the University of Uppsala and would become one of the leading chemists in Sweden and beyond [Fors, 2008, pp. 41, 47, 49]. In 1775, Scheele gained the post

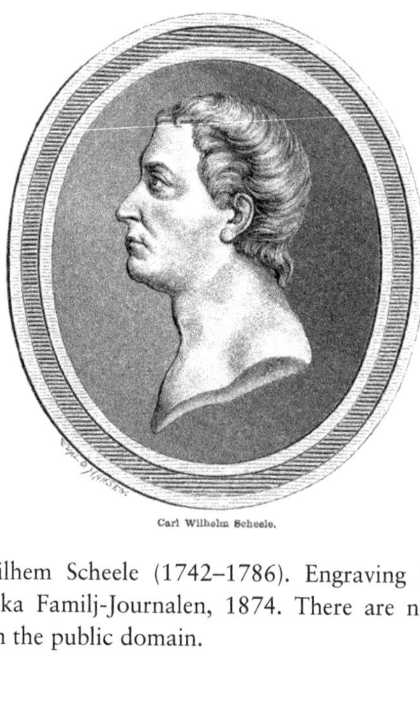

Carl Wilhelm Scheele.

Figure 4: Carl Wilhem Scheele (1742–1786). Engraving by Evald Waldemar Hansen from Svenska Familj-Journalen, 1874. There are no real contemporary images of Scheele. In the public domain.

of pharmacist in the town of Köping, where he resided until his death.

It seems that sometime in 1771, Bergman suggested to Scheele that he should study the chemistry of the mineral pyrolusite, which had been mined from the 16th century onwards for its ability to impart the colour of violet to glass and earthenware. The characteristics of pyrolusite, particularly its property of colouring glass, had led Bergman to suspect that the mineral contained a peculiar (unknown) metal [Bergman, 1970, p. 315]. By the end of 1773, Bergman received Scheele's manuscript on research he had conducted and quickly translated it into Swedish.[11] That Scheele's paper "Om Brun-sten eller Magnesia, och dess egenskaper" (On the Brown-stone or Magnesia and its Properties) was published in the *Kungliga Vetenskapsakademiens Handlingar* (Transactions of the Royal Swedish Academy of Sciences) in 1774 [Fors, 2008, pp. 40–46, 48] testifies to his standing as a skilful chemist and experimentalist and marked his acceptance by the Swedish community of chemists [Scheele, 1774].[12]

It is worth noting that the term "manganese" at the time had a different meaning than today. Scheele's term "magnesia" appeared with the same name in the English translation of his paper but with the name of "manganese" in the French version.[13] In the following, therefore, in accordance with Scheele's usage, the term "manganese" refers to the oxide rather than to the pure metal.

In his paper, Scheele described the reactions of manganese with the main acids — vegetable, animal and mineral — and other substances well known to 18th-century chemists. The paper provides

[11] Although Scheele was born in Stralsund, a town in present-day Germany, which at the time was part of the Swedish province of Pomerania, he spoke and wrote in German. This should be viewed in light of the fact that many Swedish towns had large German-speaking communities and that German culture was a common feature in the pharmaceutical profession [Fors, 2008, p. 34].

[12] For the different translations of this paper, see [Scheele, 1785; 1786; 1788; 1793; 1905; 1930].

[13] At the beginning of the 18th century, the word "magnesia" was also employed for the whitish basic carbonate of magnesium.

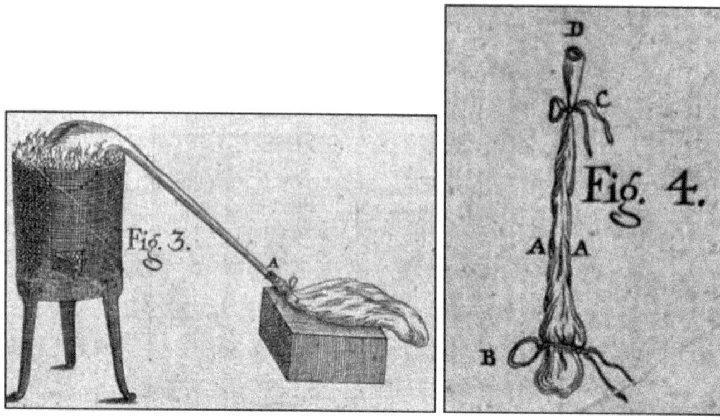


Figure 5: Apparatus and utensils used by Scheele for investigating gases. Left: Device for collecting oxygen. Right: Ox bladder for collecting gases. From the plate of his *Chemische Abhandlungen von der Luft und dem Feuer*, 1777. In the public domain.

no depiction of the experimental device he used in this experiment. Having few resources, Scheele was obliged to use rudimentary and inexpensive material for his investigation, such as that described in his *Chemische Abhandlungen von der Luft und dem Feuer*. His apparatus and vessels were the simplest that could possibly be found in a chemical laboratory compared to Priestley's equipment: matrasses, retorts, bottles, vials, and bladders (Figure 5). The apparatus and utensils were either easy to purchase or self-made; for instance, the bladders, when still fresh, were rubbed, fully inflated, bound up tightly, and hung for drying [Scheele, 1777, pp. 29–34; 1780, pp. 35–41].

After pouring muriatic acid (hydrochloric acid)[14] onto finely ground manganese, the acidic solution took on a dark brown colour.

[14] Sheele used the Latin term *spiritus salis* (spirit of salt) in his original paper. This term refers to the ancient obtention of the acid by heating rock salt (sodium chloride) and green vitriol (Iron(II) sulfate), and the name "muriatic acid" shares the same origin ("muriatic" comes from the Latin word *muriaticus*, meaning "pertaining to brine or salt"). To avoid confusion, the term *muriatic acid* will be used throughout.

Part of this solution was poured into a bottle and left open in a warm place. The solution gave off a smell like warm *aqua regis*.[15] As soon as the remainder of the brown mixture became warm, its smell of *aqua regis* became considerably augmented, and an effervescence also arose. Scheele later refined the preparation and isolation of the released gas. He took a retort containing a mixture of manganese and muriatic acid, and in front of the neck of the retort, he bound a bladder emptied of air and set the retort in hot sand. The bladder became distended by the effervescence in the retort. When the acid no longer effervesced, he removed the bladder and found that the new air had coloured it yellow, and it now had a characteristically suffocating smell resembling that of *aqua regis*. He had, in fact, isolated a new gas, which he called dephlogisticated muriatic acid [Scheele, 1930, pp. 21–22, 31–32].

Scheele thus interpreted this chemical process in accordance with the prevailing theory of phlogiston. The manganese was attracted by the muriatic acid, which released part of its phlogiston, which, in turn, was trapped by the manganese. That part of the acid that had thus lost phlogiston (and was only very loosely united to the manganese) was expelled as a previously unknown air: the dephlogisticated muriatic acid.[16] He also described many of the properties of the new acid gas, such as its reactivity and bleaching action: blue litmus paper became almost white, all vegetable flowers — red, blue and yellow — became white in a short time, and all metals were attacked. Scheele stated that dephlogisticated muriatic acid produced all these effects by subtracting phlogiston from other materials in contact with it [Scheele, 1930, pp. 31–32].

[15] *Aqua regis* or *aqua regia* is the name for a mixture of nitric and hydrochloric acids, optimally in a ratio of 1:3. It is a yellow-orange (sometimes red) fuming liquid, so-named by alchemists because it can dissolve the noble metals, gold and platinum, although not all metals.

[16] Dioxide of manganese (manganese) and hydrochloric acid (muriatic acid) react to give dichloride of manganese (dark brown solution), water and chlorine (dephlogisticated muriatic acid).

## Reinterpreting Dephlogisticated Muriatic Acid According to the Oxygen Theory

Scheele's paper of 1774 was published while the so-called chemical revolution led by Lavoisier was underway. Central to the work was the overthrow of the phlogiston theory by Lavoisier's oxygen theory. During that time, Lavoisier carried out a series of experiments on the weight gain of several substances on combustion. In June 1783, Lavoisier and the polymath Pierre-Simon Laplace (1749–1827) conducted an experiment on the synthesis of water. This constituted a crucial step that culminated in the large-scale experiments of 1785, which indicated that the hydrogen released in the action of dilute acids on metals came from the water. Lavoisier was subsequently able to arrive at a definitive formulation of the oxygen theory. According to this theory, the calcination of metals and the combustion of coal should not be understood as the release of the phlogiston contained in these materials but rather as their absorption of a chemically distinct part of common air — the vital air, later called oxygen.[17]

The chemical revolution was about more than finding a new explanation of combustion and calcination. Indeed, Lavoisier's theory of respiration was an extension of his theory of combustion, and his oxygen theory could be regarded as forming part of a general theory of the gaseous state, as well as a particular theory of acidity.[18]

A central moment in the chemical revolution was the publication in 1789 of Lavoisier's *Traité élémentaire de chimie* ("Elementary treatise on chemistry"), in which he rejected atoms as metaphysical entities as well as the existence of a few "elements" of which all

---

[17]For a general comprehension of the chemical revolution, see [Brock, 1992, pp. 87–127] and [Fauque, 2003].

[18]The prospect of the 200th anniversary in 1994 of the death of Lavoisier prompted the publication of a number of works with different approaches to his life and research. See [Bensaude-Vincent, 1993; Donovan, 1993; Poirier, 1993]. Klein [2015] and Chang [2015] have, however, challenged the idea of chemical revolution in the 18th century. Similar discussions have taken place for the so-called scientific revolution; see [Shapin, 1996].

materials were composed.[19] Instead, he introduced a number of simple bodies from an operational perspective. They would be simple insofar as they behaved as indecomposable substances in the laboratory, which left the door open for some to be decomposed later.

Lavoisier was the leader of a group of chemists who worked in his laboratory at the Arsenal in Paris, one of whom was Claude-Louis Berthollet (1748–1822), who was well known for his studies of dephlogisticated muriatic acid and its application to bleaching. His work on the development of a new bleaching process, based on the use of what we now call chlorine, made a name for him among industrial chemists.[20] The first paper that Berthollet (Figure 6)

Figure 6: Claude Louis Berthollet (1748–1822). Lithography by N. E. Maurin, 1832. Wellcome Library no. 1053i. In the public domain.

[19] Lavoisier was referring to both the four Aristotelian elements (water, air, fire and earth) and the three Paracelsian principles (salt, sulfur and mercury).

[20] Berthollet's writings on bleaching reached their final form in 1804, in the second edition of his work *Éléments de l'art de la teinture*, first published in 1791. On the development of chlorine bleaching in France, see [Smith, 1979, pp.113–191].

devoted to dephlogisticated muriatic acid was that of April 1785, in which he announced his adhesion to the Lavoisier oxygen theory [Berthollet, 1785].[21] In December of the same year, he presented to his colleagues a new study on the muriatic acid, developing his argumentation and observations. His conclusion was that dephlogisticated muriatic acid was not the result of a loss of phlogiston from muriatic acid but of the combination of the latter with the vital air (oxygen) contained in manganese [Berthollet, 1788]. A property that attracted Berthollet's attention was that dephlogisticated muriatic acid caused no effervescence with alkaline solutions[22] and did not dissolve iron and zinc with either water decomposition or inflammable air (hydrogen) release,[23] a behaviour unlike any other known acids when reacting with some alkalis and metals.

As mentioned above, Lavoisier's oxygen theory was not only a theory on the calcination of metals and combustion but also a theory of acidity. Because of the fact that sulfur, phosphorus and carbon all formed acids with an excess of oxygen, Lavoisier generalised that all acids contain oxygen. The case of dephlogisticated muriatic acid was regarded as an exception. Indeed, according to Lavoisier's theory of acidity, muriatic acid that supposedly already contained oxygen could take up an excess of oxygen to form the new gas. The acidity of this exceedingly oxygenated substance appeared to diminish rather than increase. Although such a counter example defied the new theory of acidity, the balance of evidence still remained in its favour.

The promoters of the new chemical nomenclature agreed that dephlogisticated muriatic acid required a new denomination to match its particular characteristics; thus, it was renamed oxygenated

---

[21] Berthollet was the first prominent chemist to align himself with Lavoisier against the followers of phlogiston.

[22] Chlorine reacts with dilute alkaline solutions to give a mixture of alkaline hypochlorite and chloride, the former changing to chlorate by heating.

[23] Chlorine usually does not have an effect on metals such as iron, copper, platinum, and silver at temperatures below 110°C. At higher temperatures, however, it reacts rapidly with many of the metals and forms metal chlorides.

muriatic acid [Guyton de Morveau *et al.*, 1787, pp. 45–46]. The Irish chemist Richard Kirwan (1733–1812) referred to it as oxymuriatic acid, a term used in English-speaking countries.

In 1806, jointly with Laplace, Berthollet set up a private scientific society, the *Société d'Arcueil* (Society of Arcueil), which brought together a select group of young scientists.[24] Two of these scientists later recognised oxygenated muriatic acid as a simple body. As a research supervisor of the Arcueil group, Berthollet was, however, to play a controversial role in that matter.

## The Simple Nature of Oxymuriatic Acid: A Deferred Discovery

The two young scientists were Joseph Louis Gay-Lussac (1778–1850) and Louis Jacques Thénard (1777–1857). Gay-Lussac had been appointed to the post of demonstrator at the *École Polytechnique* in 1804, where he had attended the chemistry lectures of Louis Bernard Guyton de Morveau (1737–1816) and Berthollet. Gay-Lussac had the good fortune to be recruited for the Arcueil group by Berthollet, who eventually became the supervisor of his scientific career. After that, Gay-Lussac divided his time between the *École Polytechnique* and the Arcueil circle.

As for Thénard, he came to Paris with the intention of training to become a pharmacist. Louis Nicolas Vauquelin (1763–1829), one of his teachers, found him a job as a private teacher and later as a demonstrator at the *École Polytechnique*. After 1804, Thénard, who at that time taught chemistry at the *Collège de France*, became a member of the Arcueil group, and his most important research in this milieu was conducted in collaboration with Gay-Lussac [Crosland, 1978, pp. 71–73].

[24] The role of the *Société d'Arcueil* in the establishment of French science after Lavoisier's death has been exhaustively studied by Crosland [1967].

Figure 7: Joseph Louis Gay-Lussac (1778–1850) and Louis Jacques Thénard (1777–1857). Engraving of Gay-Lussac by C. W. Sharpe after Zéphirin Belliard. Wellcome Library no. 3461i. In the public domain. Lithography of Thénard by Julien Léopold Boilly. Wellcome Library no. 9134i. In the public domain.

In February 1809, Gay-Lussac and Thénard (Figure 7) read a paper on oxygenated muriatic acid and muriates at the *Institut de France*. The day before, in a meeting at the Society of Arcueil, they had presented an initial version of this paper in which, without affirming it categorically, they suggested with some confidence that oxygenated muriatic acid was a simple body.

Two decisive experiments prompted them to believe that the constitution of oxygenated muriatic gas[25] was quite different than originally thought. Firstly, they tried to decompose oxygenated muriatic gas by charcoal ignited in the extreme heat of a forge. The expected outcome was that oxygen would be extracted from oxygenated muriatic gas by charcoal, yielding muriatic acid and carbonic oxide gas (carbon monoxide). Secondly, they assembled an apparatus that enabled them to introduce oxygenated muriatic gas into a porcelain tube exposed to a red heat. Lavoisier had previously set up a similar experimental procedure with the aim of proving that water

[25] Oxygenated muriatic gas referred to pure chlorine gas, whereas oxygenated muriatic acid referred to the aqueous solution of chlorine.

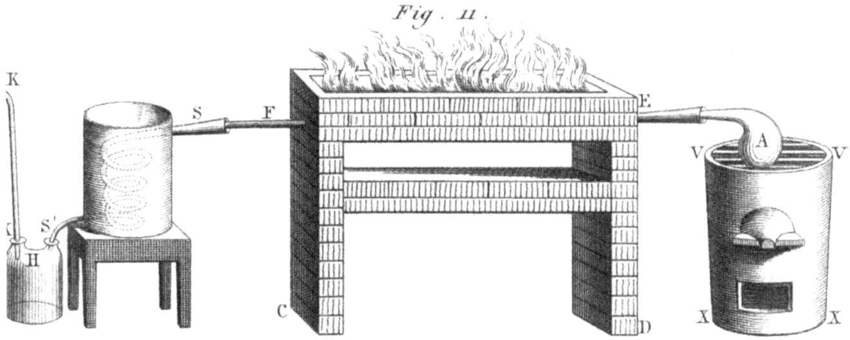


Figure 8: Lavoisier's apparatus to prove the compound nature of water. From *Traité élémentaire de chimie* (Paris, 1789), vol. 2, plate VII. In the public domain.

was a compound substance (Figure 8). At that time, however, he used a heat-resistant glass tube instead of a porcelain one because of the difficulty involved in procuring a tube so entirely free from pores as to prevent the leakage of vapours [Lavoisier, 1789, 1, pp. 87–96].

Oxygenated muriatic gas resisted the action of these most energetic agents, and it was never decomposed. Actually, it was only decomposed in the presence of water or hydrogen, a finding that led the two young chemists to the conclusion that muriatic acid gas could not exist without the concomitant presence of water.[26]

Berthollet, who had also studied these properties, found Gay-Lussac and Thénard's assumption to be too audacious and persuaded them to be very cautious in making that assertion. That oxygenated muriatic gas contained oxygen was evident to Berthollet, who had indicated that the aqueous solutions of the gas exposed to sunlight released oxygen.[27] Berthollet's influence was so strong that, in the

[26]They also found that a mixture, in equal parts, of oxygenated muriatic gas and hydrogen gas changed into ordinary muriatic gas over the course of several days and that no water was yielded.

[27]The addition of chlorine to water gives both hydrochloric acid and hypochlorous acid (in accordance with comment to note 23). The latter partially dissociates in water, giving a hypochlorite anion. Both species (hypochlorous acid and hypochlorite anion) undergo photolysis by sunlight and release oxygen, hydrochloric acid, and chloride anion.

definitive version of the paper presented the next day, both followed his advice and confined themselves to suggesting that the simple nature of oxygenated muriatic acid was a suitable hypothesis for interpreting its properties but without going so far as to embrace this hypothesis.

> In fact, oxygenated muriatic acid is not decomposed by charcoal, and from this fact and those that are communicated in this Memoir, it might be supposed that this gas is a simple body. The phenomena that it presents can be explained well enough by this hypothesis; we shall not seek to defend it, however, as it appears to us that they are still better explained by regarding oxygenated muriatic acid as a compound body.
>
> [Gay-Lussac and Thénard, 1809, p. 357–358; 1905, p. 48].

While agreeing to modify his original conclusions for the publication, Gay-Lussac was already teaching the new approach to the elementary nature of oxygenated muriatic acid in his lectures of 1809. Probably because of the pressure Berthollet exerted on Gay-Lussac and Thénard in the environment of the Society of Arcueil, which was basically perpetuating Lavoisier's chemistry, the discovery of the elementary nature of chlorine is often credited to Davy, who revealed it publicly in 1810. Yet, the full story is anything but simple and requires turning the page back a few years [Sadoun-Goupil, 1977, pp. 81–82; Crosland, 1967, pp. 129, 309–310].

## The Acceptance of Chlorine as a New Simple Body

In December 1794, Davy's family moved to Penzance in the South of England, where Humphrey was apprenticed to a local apothecary surgeon. In late 1797, Davy began to study chemistry using Lavoisier's (translated) textbook *Elements of Chemistry* and also apparently William Nicholson's *Dictionary of Chemistry*. Davy took up the post of assistant at the Pneumatic Institution in Clifton (a suburb of Bristol) in 1798. His careful experimental work on the isolation and the chemical and physiological properties of the various oxides of nitrogen was published in 1800 in his monograph *Researches Chemical and Philosophical*, mainly concerning nitrous oxide [Davy, 1800]. In March 1801, Davy gave up his post of superintendent at the Pneumatic Institution and moved to the Royal Institution in London, where he was appointed as an assistant lecturer in chemistry.

In 1800, the announcement was made of the invention of an instrument that opened up to chemists a new territory that would become quite fruitful in the practical exploration, theoretical innovation and discovery of new elements. Alessandro Volta (1745–1827) revealed the invention of an electric pile by means of a letter sent to the secretary of the Royal Society of London. The letter was first read on 26 June 1800 and published the following September in the *Philosophical Transactions,* although prior to its publication, the contents of that letter were already known by Anthony Carlisle (1768–1840) and William Nicholson (1753–1815). Carlisle was a distinguished physician at Westminster Hospital in London and Nicholson was a renowned chemist who served as an engineer, inventor, translator, and scientific publisher. In May 1800, they began to make their own pile with which they were able to decompose water. The following year, in the laboratory of the Royal Institution, Davy (Figure 9) began an accelerated research project to take advantage of the analytical potential of Volta's pile. Thus, in October 1807, by building a powerful electric battery and using fused electrolytes, he was able to announce the decomposition of soda and potash and the isolation of two new elementary metals, sodium and potassium. Over the next few years, he prepared calcium, strontium and barium electrolytically.

In Davy's hands, the electric battery became a powerful analytical instrument, and in his mind, it also generated a new conception of chemical affinity by replacing Newtonian attraction with forces of an electrical nature.[28] Davy's contributions, along with those of Hans Christian Oersted (1777–1851), André Marie Ampère

---

[28] See Footnote 4.

Figure 9: Humphry Davy (1788–1829). Engraving by S. Freeman after J. Londsdale, 1822. Wellcome Collection. In the public domain.

(1775–1836), and Michael Faraday (1791–1867), constituted the first attempts to construct an electrochemical theory of chemical change that would eventually be brought to fruition by (Jöns) Jacob Berzelius (1779–1848) by 1811 [Levere, 1971, pp. 23–67, 107–139; Sadoun Goupil, 1991, pp. 207–222].

The French chemists of the Arcueil circle and Davy knew each other quite well. The story of Davy's chemical research in London and Gay-Lussac and Thénard's in Paris from 1807 onward is one of intense rivalry ignited by Davy's isolation of sodium and potassium.

Motivated by Davy's work, Gay-Lussac and Thénard decided to join forces. They succeeded, for instance, in obtaining potassium and sodium by an alternative method that consisted of heating the corresponding alkali at high temperatures with iron filings. The advantage of this method was that it enabled those metals to be obtained in larger quantities, and hence, at a lower cost [Crosland, 1978, p. 75]. This confirmed both Davy's results and the use of the voltaic pile as a legitimate means of decomposing compound bodies. The pile had thus acquired the status of an instrument providing evidence that was considered as acceptable as the analytical chemical processes established by pneumatic and antiphlogistic chemistry. Although Davy was able to consolidate the electric pile as a reliable instrument of analysis, agreement on the interpretation of the effects of the electric pile was not unanimous. Thus, for instance, Davy's findings could, at that time, have been regarded as a resumption of the phlogiston theory and the elementary nature of alkaline metals open to question [Golinski, 1992, pp. 213–214].

The French chemists, in turn, inspired Davy. In a paper read to the Royal Society in July 1810, he commented on the important 1809 paper by Gay-Lussac and Thénard described in the previous section:

> One of the most singular facts that I have observed on this subject, and which I have before referred to, is that charcoal, even when ignited to whiteness in oxymuriatic or muriatic acid gases by the Voltaic battery, effects no change in them; if it has been previously freed from hydrogen and moisture by intense ignition in vacuum. This experiment, which I have several times repeated, led me to doubt of the existence of oxygen in that substance, which has been supposed to contain it above all others in a loose and active state; and to make a more vigorous investigation than had been hitherto attempted for its detection.
>
> [Davy, 1810, p. 232; 1840, p. 285; 1894, pp. 22–23]

Davy was particularly impressed by the fact that even at white heat, charcoal could not trigger the decomposition of oxygenated muriatic gas. He was also able to confirm that no decomposition occurred when employing a powerful voltaic battery of 1,000 double plates [Davy, 1810, p. 239–240; 1840, p. 292–293; 1894, pp. 30]. He ultimately suggested that oxygenated muriatic gas might be a simple body, and in November 1810, he proposed that it should be named chlorine:

> To call a body which is not known to contain oxygen, and which cannot contain muriatic acid, oxymuriatic acid, is contrary to the principles of that nomenclature in which it is adopted; and an alteration of it seems necessary to assist the progress of discussion, and to diffuse just ideas on the subject. ... dephlogisticated marine acid is a term which can hardly be adopted in the present advanced era of the science. After consulting some of the most eminent chemical philosophers in this country, it has been judged most proper to suggest a name founded upon one of its obvious and characteristic properties — its colour and to call it *Chlorine*, or *Chloric* gas.[29]
>
> [Davy, 1811, p. 32; 1840, p. 345; 1894; pp. 59–60]

That same year, Gay-Lussac and Thénard still confirmed that they viewed oxygenated muriatic acid as a compound [Gay-Lussac and Thénard, 1811, pp. 171–172, 262].[30] And, in 1814, Gay-Lussac mentioned with some regret the circumstances in which this priority had eluded both him and Thénard [Gay-Lussac, 1814, pp. 96–97]. This demonstrates that the community of chemists was well aware that this indecomposability of oxymuriatic gas was considered an important discovery worthy of credit.

For his part, Berthollet remained determinedly convinced that chlorine was an oxygenated substance [Berthollet *et al.*, 1811, pp. 126–128]. In Berthollet's defence, it should be remembered that some leading chemists of the day, including John Dalton (1766–1844) and Berzelius, initially declined to accept Davy's plausible evidence for the elementary nature of chlorine. Even Thénard commented that despite all the phenomena produced by oxygenated muriatic gas, it could be explained either by considering it as a simple or as a compound body, although he preferred the latter

[29] From the Greek word χλωρος (*chlōros*, "green-yellow"), in reference to its colour. Emphasis in the original.

[30] This endorsement reactivated the French chemist-pharmacist François René Curaudau's (1765–1813) priority claim made in 1810 on the topic of the elementary nature of chlorine [Chabot, 2006, pp. 139–162].

hypothesis [Thénard, 1813–1816, 1, p. 562n]. In 1816, Berthollet finally admitted that it was due to his interposition that Gay-Lussac and Thénard had desisted from publishing their opinion in 1809 [Sadoun-Goupil, 1977, pp. 134–135]. It was at that time that Berthollet changed his mind about regarding oxygenated muriatic acid as a simple body:

> [...] it seems to me that we can find proofs within the presently known facts, which oblige us to consider chlorine as a simple substance, and similar to all those we contemplate as elementary.[31]
>
> [Berthollet, 1817, pp. 604–605]

Berzelius' opposition to the acceptance of the elementary nature of chlorine was an issue of lasting persistence. One of the reasons for his opposition was that such an acceptance came into conflict with his conception of oxidation reflected in his rule of oxides.[32] According to Berzelius, the number of the saline combinations of muriatic acid proved that it contained oxygen. Another clue to understanding his resistance was that the compound nature of chlorine fitted better with his electrochemical theory. Nevertheless, the fact is that it was not until 1826 that Berzelius listed chlorine as an elementary substance [Gray *et al.*, 2007, pp. 60–67]. The acceptance of chlorine as a simple substance was probably due to the recognition of its analogous chemical properties to iodine and the consequent growing evidence for its elementary nature rather than to a strong consensus based on the previously conducted experiments by Gay-Lussac, Thénard and Davy.

As in the case of chlorine, the assertion of the elementary nature of iodine was a quite complex story that began with the fortuitous finding of Bernard Courtois (1777–1838) — a saltpetre manufacturer — who in 1811 observed the formation of a violet

[31] This translation into English is by the author.

[32] "In neutral salts the amount of oxygen contained by the acid is an integral multiple of the amount of oxygen in the base" [Berzelius, 1812, p. 304]. English translation from Melhado [1992, p. 148].

vapour when treating a kind of marine algae (kelp) with sulfuric acid. This vapour, once condensed, yielded a small amount of black crystals. The rest of the story is largely a return to the rivalry between Davy and Gay-Lussac, both of whom began their research on the new substance quite independently, although it was Gay-Lussac who first read an account of his experiments on 6 December 1813, to the First Class of the *Institut de France*, in which he suggested that the new substance should be called "iode".[33] To secure priority of publication, Gay-Lussac published the corresponding paper in the *Moniteur* (the government newspaper) on 12 December, while Davy's first public account was given on the following day at a meeting of the First Class. Both emphasised the resemblance of the new substance to chlorine, and Gay-Lussac recognised its elementary nature. The similarities between chlorine and iodine were clearly set out in Gay-Lussac's complete study on iodine published in 1814 [Siegfried, 1959; Crosland, 1978, pp. 80–91].

Chemists who remained unconvinced about the elementary nature of chlorine were only persuaded to accept it when it began to be perceived that iodine, once recognised as a new elementary substance, had chlorine-like properties. By 1816–1818, most chemists had gradually come to accept chlorine as a new simple substance [Hijmans and Llored, 2020, pp. 449].

* * *

The history of the discovery of most elements cannot be reduced to the statement that a chemical element has a unique "discoverer", nor that the discovery can be pinpointed to a specific moment, and chlorine is certainly no exception. In this contribution, I have focused on two stages of the process that took more than 40 years, from the moment when Scheele identified a new air by describing its characteristic properties (stage 1) to the point that leading chemists agreed to view this substance as simple, and further indecomposable by means of chemical analysis (stage 2). The isolation of chlorine and

[33] From the Greek word *ἰώδης* (violet-coloured), in reference to its violet colour.

the recognition of its elementary nature was a process that involved different actors operating in distinct contexts, both theoretical and experimental.

It was in the context of the Swedish mining tradition that Bergman put a sample of pyrolusite into the hands of Scheele for examination. This constituted the beginning; mining interests provided the motivation for Bergman to study one mineral rather than another. Scheele was able to isolate a new kind of air using the well-established methods of recent pneumatic chemistry; this gaseous substance had to be described and was characterised as a compound within the framework of the phlogiston theory. The emergence of the oxygen theory would provide chemists with an operational concept of a simple body that could be checked in a laboratory by means of a balance, as well as an improved version of experimental devices for investigating the nature of gaseous substances. Whereas the oxygen theory regarded Scheele's air as a compound (called oxymuriatic acid), the definition of a simple substance and its articulation with experimental practice enabled French chemists Gay-Lussac and Thénard to argue for the elementary nature of chlorine on the basis of their own experimental research. However, together with the pressure from their patron Berthollet, the fact that the results of their investigation could, to some extent, still be construed with oxygenated muriatic acid as a compound body led them to refrain from upholding their former assumption in favour of the commonly preferred oxygen theory of acidity.

The debate on the elementary nature of chlorine crossed the Channel when Davy presented oxymuriatic air unequivocally as a simple body. He did so by making appropriate use of the earlier research by the French chemists and without introducing any decisive experimental novelty, except for the use of the voltaic pile, which turned out to be decisive for the isolation of sodium, potassium, calcium, strontium, and barium. This development placed the dispute squarely in the context of the ongoing rivalry between Davy and Gay-Lussac.

However, chlorine definitively gained the full status of a simple substance only when its chemical resemblance with iodine was recognised. In fact, chemical behaviour had similarly been already used as an organising principle in late 18$^{th}$-century chemistry (e.g., combustible and non-combustible bodies). Furthermore, at the beginning of the next century, chemical similarities were also employed for grouping together analogous bodies. Thus, Thénard ordered substances by their affinity for oxygen [Bertomeu-Sánchez *et al.*, 2002, pp. 234–237]. Consideration of a chemical body as a candidate for a simple substance should thus also be framed within the wider context of its classification, along with other elementary substances with similar chemical behaviour [Hijmans and Llored, 2020, pp. 453–454].

Notwithstanding, the debate on chlorine's elementary nature has provided the opportunity to review the indecomposability of a substance as one of the criteria for its classification as a chemical element. Providing that evidence was the work of several chemists, practising their science in a variety of sites, with their own hierarchy and processes of dissemination, and entrenched in successive conceptual and experimental frameworks. In that regard, it is impossible to isolate the process from the context.

## Acknowledgements

The author wishes to thank the editors for their careful editorial work on this chapter, and Brigitte Van Tiggelen, in particular, for her valuable comments and suggestions on an earlier draft of this chapter.

## References

Bensaude-Vincent, B. (1993). *Lavoisier. Mémoires d'une Révolution* (Flammarion, Paris).

Bensaude-Vincent, B. and Stengers, I. (1993). *Histoire de la chimie* (Éditions La Découverte, Paris).

Bergman, T. O. (1970). *A Dissertation on Elective Attractions*. Translated by Thomas Beddoes (1785) from the original Latin edition of 1775 and with a new introduction by A. M. Duncan (Frank Cass & Co., London).

Berthollet, C. L. (1785). Mémoire sur l'acide marin déphlogistiqué, *Observations sur la physique, sur l'histoire naturelle et sur les arts*, **26**, pp. 321–330.

Berthollet, C. L. (1788). Mémoire sur l'acide marin déphlogistiqué, *Histoire de l'Académie Royale des Sciences,* pour l'année 1785, pp. 278–288.

Berthollet, C. L. (1817). Note sur la décomposition de l'acide oximuriatique, *Mémoires de physique et de chimie de la Société d'Arcueil* (Chez M$^{me}$. V$^{e}$. H. Perronneau, Paris), **3**, pp. 603–611.

Berthollet, C. L., Chaptal, J. A. and Vauquelin, L. N. (1811). Suite du rapport sur un mémoire de M. Curaudau, *Annales de chimie,* **80**, deuxième cahier, pp. 121–139.

Bertomeu-Sánchez, J. R., García-Belmar, A. and Bensaude-Vincent, B. (2002). Looking for an order of things: Textbooks and chemical classifications in nineteenth century France, *Ambix*, **49**, 3, pp. 227–250.

Berzelius, J. J. (1812). Dritte Fortsetzung des Versuch, die bestimmten und einfachen Verhältnisse aufzufinden, nach welchen die Bestandtheile der unorganischen Natur mit einander verbunden sind, *Annalen der Physik*, **40**, pp. 235–330.

Black, J. (1756). Experiments on magnesia alba, quicklime and some other alcaline substances, *Essays and Observations, Physical and Literary*, **2**, pp. 157–225; (1898) [Alembic Club Reprints, 1] (William F. Clay, Edinburgh; Simpkin *et al.*, London) (http://www.gutenberg.org/files/24591/24591-h/24591-h.htm, last accessed 7 January 2021).

Blumenthal, G. and Ladyman, J. (2017). The Development of problems within the phlogiston theories, 1761–1791, *Foundations of Chemistry*, **19**, pp. 241–280.

Brock, W. H. (1992). *The Fontana History of Chemistry* (Fontana Press, London).

Cavendish, H. (1766). Three papers, containing experiments on factitious air, *Philosophical Transactions*, **46**, pp. 141–183.

Chabot, H. (2006). Les origines d'un nouvel élément chimique: l'affaire du chlore. In *Du nouveau dans les sciences.* [Recherches sur la philosophie et le langage, 24], eds. Carvallo, S. and Roux, S. (Université Pierre Mendès France, Département de philosophie, Grenoble), pp. 121–170.

Chang, H. (2015). The chemical revolution revisited, *Studies in History and Philosophy of Science*, **49**, pp. 91–98.

Crosland, M. (1967). *The Society of Arcueil. A View of French Science at the Time of Napoleon I* (Harvard University Press, Cambridge, MA, USA).

Crosland, M. (1978). *Gay-Lussac, Scientist and Bourgeois* (Cambridge University Press, Cambridge, London).

Crosland, M. (2000). "Slippery substances": Some practical and conceptual problems in the understanding of gases in the pre-Lavoisier era. In *Instruments and Expermentation in the History of Chemistry*, eds. Holmes, F. L. and Levere, T. H. (The MIT Press, Cambridge, MA, USA and London), pp. 79–104.

Davy, H. (1800). *Researches Chemical and Philosophical, Chiefly Concerning Nitrous Oxide or Dephlogisticated Nitrous Air and its Respiration* (J. Johnson, London; Biggs and Cottle, Bristol).

Davy, H. (1810). Researches on the oxymuriatic acid, its nature and combinations; and on the elements of the muriatic acid. With some experiments on sulphur and phosphorus, made in the laboratory of the Royal Institution, *Philosophical Transactions of the Royal Society of London*, **100**, pp. 231–257; (1840). *The Collected Works of Sir Humphry Davy*, ed. Davy, J. (Smit *et al.*, London), 9 vols., 5, pp. 284–311; (1894). *The Elementary Nature of Chlorine. Papers by Humphry Davy (1809–1818)* [Alembic Club Reprints, 9] (William F. Clay, Edinburgh; Simpkin *et al.*, London), pp. 21–39.

Davy, H. (1811), The Bakerian Lecture. On some of the combinations of oxymuriatic gas and oxygen and on the chemical relations of these principles to inflammable bodies, *Philosophical Transactions of the Royal Society of London*, **101**, pp. 1–35; (1840) *The Collected Works of Sir Humphry Davy*, ed. Davy, J. (Smit *et al.*, London), 9 vols., 5, pp. 312–348; (1894), *The Elementary Nature of Chlorine. Papers by Humphry Davy (1809–1818)* [Alembic Club Reprints, 9] (William F. Clay, Edinburgh; Simpkin *et al.*, London), pp. 40–62.

Donovan, A. (1993). *Antoine Lavoisier. Science, Administration and Revolution* (Blackwell, Oxford).

Donovan, A. L (1975). *Philosophical Chemistry in the Scottish Enlightenment. The Doctrines and Discoveries of William Cullen and Joseph Black* (Edinburgh University Press, Edinburgh).

Duncan, A. (1996). *Laws and Order in Eighteenth-Century Chemistry* (Clarendon Press, Oxford).

Fauque, D. (2003). *Lavoisier. La naissance de la chimie moderne* (Vuibert, Paris).

Fors, H. (2003). *Mutual Favours. The Social and Scientific Practice of Eighteenth-Century Swedish Chemistry* (Universitetstryckeriet, Uppsala).

Fors, H. (2008). Stepping through science's door: C.W. Scheele, from pharmacist's apprentice to man of science, *Ambix*, **55**, 1, pp. 29–49.

Gay-Lussac, J. L. (1814). Mémoire sur l'Iode, *Annales de chimie*, **91**, pp. 5–160.

Gay-Lussac, J. L. and Thénard, L. J. (1809), Extrait des mémoires lus a l'Institut National depuis le 7 mars 1808 jusqu'au 27 février 1809: De la nature et des propriétés de l'acide muriatique et de l'acide muriatique oxygéné, *Mémoires de Physique et de Chimie de la Société d'Arcueil.* (Mad. V$^{e}$. Bernard, Paris), **2**, pp. 339–358; (1905). *The Early History of Chlorine* [Alembic Club Reprints, 13] (William F. Clay, Edinburgh; Simpkin *et al.*, London), pp. 37–48.

Gay-Lussac, J. L. and Thénard, L. J. (1811), *Recherches physico-chimiques* (Deterville, Paris), 2 vols.

Golinski, J. (1992). *Science as Public Culture. Chemistry and Enlightenment in Britain, 1760–1820* (Cambridge University Press, Cambridge, MA, USA), pp. 225–232.

Gray, T., Coates, R. and Åkesson, M. (2007). The elementary nature of chlorine. In *An Element of Controversy. The Life of Chlorine in Science, Medicine, Technology and* War [BSHS Monographs, 13], eds. Chang, H. and Jackson, C. (BSHS, n. p.), pp. 41–72.

Guyton de Morveau, L. B., Lavoisier, A. L., Berthollet, C. L. and Fourcroy, A. F. (1787). *Méthode de nomenclature chimique* (Cuchet, Paris).

Hijmans, S. and Llored, J. P. (2020). How to investigate the underpinnings of sciences? The case of the element chlorine, *Foundations of Chemistry*, **22**, pp. 447–456.

Ihde, A. L. (1984). *The Development of Modern Chemistry* (Dover Publications, New York, NY, USA). Originally published in 1964 (Harper & Row, New York, NY, USA).

Klein, U. (2015). A revolution that never happened, *Studies in History and Philosophy of Science*, **49**, pp. 80–90.

Lavoisier, A. L. (1789). *Traité élémentaire de chimie* (Cuchet, Paris), 2 vols.

Levere, T. H. (1971). *Affinity and Matter. Elements of Chemical Philosophy 1800–1865* (Clarendon Press, Oxford).

Melhado, E. M. (1992). Novelty and tradition in the chemistry of Berzelius. In *Enlightenment Science in the Romantic Era. The Chemistry of Berzelius and Its Cultural Setting*, eds. Melhado, E. M. and Frängsmyr, T. (Cambridge University Press, Cambridge, MA, USA), pp. 132–170.

Poirier, J. P. (1993). *Antoine Laurent Lavoisier, 1743–1749* (Pygmalion, Paris). English translation: (1996) *Lavoisier, Chemist, Biologist, Economist* (Pennsylvania University Press, Philadelphia, PA, USA).

Priestley, J. (1775). *Experiments and Observations of Different Kinds of Air* (Birmingham, J. Johnson).

Sadoun-Goupil, M. (1977). *Le chimiste Claude-Louis Berthollet (1748–1822). Sa vie, son œuvre* (J. Vrin, Paris).

Sadoun-Goupil, M. (1991). *Du Flou au Clair? Histoire de l'affinité chimique de Cardan à Prigogine* (Éditions du CTHS, Paris).

Scheele, C. W. (1774). Om Brun-sten eller Magnesia, och dess egenskaper, *Kungliga Vetenskapsakademiens Handlingar,* **35**, pp. 89–116, 177–194.

Scheele, C. W. (1777). *Chemische Abhandlungen von der Luft und dem Feur* (Verglegt von Magn. Swederus Buchländler, Upsala und Leipzig),

Scheele, C. W. (1780). *Chemical Observations and Experiments on Air and Fire* (J. Johnson, London).

Scheele, C. W. (1785). De la manganèse et de ses propriétés. In *Mémoires de Chymie*. (L'Éditeur *et al.*, Paris), 2 vols.,1, pp. 39–115.

Scheele, C. W. (1786). On manganese, magnesium, or magnesia vitrariorum. In *The Chemical Essays of Charles-William Scheele* (J. Murray, London; W. Goldon and C. Elliot, Edinburgh), pp. 67–142. Re-issued in 1901 by Scott and Greenwood & Co, London.

Scheele, C. W. (1788). De magnesia nigra. In *Opuscula Chemica et Physica*, eds. Shaffer, G. H. and Hebenstreit, E. B. G. (Io. Godofr. Mülleriana, Lipsiae), 2 vols.,**1**, pp. 227–281.

Scheele, C. W. (1793). Von Braunstein oder magnesium und von deren eigenschaften. In *Carl Wilhelm Scheele, Sämmtliche Physische und Chemische Werke*, ed. Hermbstädt, S. F. (n.p., Berlin), 2 vols., **2**, pp. 35–92.

Scheele, C. W. (1905). On manganese and its properties. In *The Early History of Chlorine* [Alembic Club Reprints, 13] (William F. Clay, Edinburgh; Simpkin *et al.*, London), pp. 5–10.

Scheele, C. W. (1930). On manganese or magnesia; and its properties. In *The Collected Papers of Carl Wilhelm Scheele* (G. Bell & Sons Ltd., London), pp. 17–49.
Schofield, R. E. (2004). *The Enlightened Joseph Priestley. A Study of His Life and Work from 1773 to 1804* (The Pennsylvania State University Press, University Park, PA, USA).
Shapin, S. (1996). *The Scientific Revolution* (The University of Chicago Press, Chicago, IL, USA and London).
Siegfried, R. (1959). Humphry Davy and the elementary nature of chlorine, *Journal of Chemical Education*, **36**, 11, pp. 568–570.
Smith, J. G. (1979). *The Origins and Early Developments of the Heavy Chemical Industry in France* (Clarendon Press, Oxford).
Thénard, L. J. (1813–1816). *Traité de chimie élémentaire, théorique et pratique*, (Crochard, Paris), 4 vols.

## Selected Bibliography

Ashbee, R. (2007). The discovery of chlorine: A window on the chemical revolution. In *An Element of Controversy. The Life of Chlorine in Science, Medicine, Technology and* War [BSHS Monographs, 13], eds. Chang, H. and Jackson, C. (BSHS, n. p.), pp. 15–40.
Berthollet, C. L., Gay-Lussac, J. L., Guyton de Morveau, L. B., Scheele, C. W. and Thénard, L. J. (1905). *The Early History of Chlorine* [Alembic Club Reprints, 13] (William F. Clay, Edinburgh; Simpkin *et al.*, London), pp. 5–10. (https://archive.org/search.php?query=The%20Early%20History%20of%20Chlorine, last accessed 7 January 2021).
Black, J. (1756). Experiments on magnesia alba, quicklime and some other alcaline substances, *Essays and Observations, Physical and Literary*, 2, pp. 157–225; (1898). [Alembic Club Reprints, 1] (William F. Clay, Edinburgh; Simpkin *et al.*, London) (http://www.gutenberg.org/files/24591/24591-h/24591-h.htm, last accessed 7 January 2021).
Brock, W. H. (1992). *The Fontana History of Chemistry* (Fontana Press, London), pp. 97–111, 147–149.
Coley, N. (1991). The chemical revolution. In *The Rise of Scientific Europe; 1500–1800,* eds. Goodman, D. and Russell, C. A. (Hodder & Stoughton Limited, Sevenoaks), pp. 387–413.

Davy, H. (1894). *The Elementary Nature of Chlorine. Papers by Humphry Davy (1809–1818)* [Alembic Club Reprints, 9] (William F. Clay, Edinburgh; Simpkin *et al.*, London), pp. 21–39. (https://archive.org/search.php?query=The%20Elementary%20Nature%20of%20Chlorine, last accessed 7 January 2021).

Gray, T. Coates, R. and Åkesson, M. (2007). The elementary nature of chlorine. In *An Element of Controversy. The Life of Chlorine in Science, Medicine, Technology and* War [BSHS Monographs, 13], eds. Chang, H. and Jackson, C. (BSHS, n. p.), pp. 41–72.

https://doi.org/10.1142/9789819813858_0004

# THE MANY FATHERS OF ALUMINIUM: COLLABORATION, CREDIT AND THE CONSTRUCTION OF DISCOVERY

Sarah N. Hijmans

*Université Paris Cité - Laboratoire SPHERE UMR 7219*
*27 rue Jean Antoine de Baïf*
*Case courier 7093*
*725205 Paris Cedex 13*
*sarah.hijmans@u-paris.fr*

On the first page of his monograph on aluminium (Figure 1), Joseph W. Richards [1896] states that his work is dedicated to Friedrich Wöhler (1800–1882) and Henri Sainte-Claire Deville (1818–1881). Richards, who wrote the book hoping it would help aluminium "[rank] next to iron in its usefulness to mankind", wanted to pay tribute to the men he saw as the fathers of aluminium: Wöhler, who discovered the metal by producing a small quantity of it in 1827, and Sainte-Claire Deville, who enabled its production on an industrial scale in 1854.

However, not everyone would agree with Richards' version of this story. According to James R. Partington [1964, p. 323], for

TO THE MEMORY OF

**Frederick Wöhler**

OF

GÖTTINGEN,

THE DISCOVERER OF ALUMINIUM,

AND OF

**Henri Sainte-Claire Deville**

OF

PARIS,

FOUNDER OF THE ALUMINIUM INDUSTRY.

THE FRUITS OF WHOSE LABORS
HAVE ENRICHED SCIENCE AND BENEFITED MANKIND.

Figure 1: Dedication of the monograph on aluminium by Joseph Richards (1896). In the public domain.

example, the Danish chemist Hans Christian Ørsted (1777–1851) "should be regarded as an earlier discoverer" for his work on the metal in 1825. Edmond Rancke-Madsen [1976, p. 308] dates the discovery even further back to 1754 when the Prussian chemist Andreas Sigismund Marggraf (1709–1782) identified the compound alumina, which contained aluminium, as a distinct chemical substance. In other words, in accounts of the history of aluminium, its discovery date varies between 1754 and 1854, depending on how one views the discovery of an element.

The question is further complicated by the fact that the discovery of aluminium is generally distinguished from the proof of its existence. Aluminium was seen as an element long before its isolation, following the work of Humphry Davy (1778–1829) on the decomposition of the earths in 1808. Nevertheless, in traditional histories of aluminium, Davy's experiments are referred to as "failed" or "unsuccessful" and excluded from the discovery process [Holmes, 1930; Jagnaux, 1891; Weeks, 1956].

How should this contradictory information be interpreted? Why is it so hard to answer the question, "Who discovered aluminium"? The main reason for the difficulty of dating the discovery of aluminium is not that this particular story is exceptionally complicated. In fact, the notion of discovery itself is problematic. Scientific discoveries have traditionally been taken for granted as clearly identifiable events carried out by one individual at a specific time. However, detailed historical analyses of cases like these show that it is impossible to pinpoint when exactly they took place and by whom — in reality, discoveries are intricate processes that extend over time and often involve multiple people.

In the words of Thomas Kuhn [1962, p. 762], discoveries of a phenomenon are necessarily complex because they involve "recognising both *that* something is and *what* it is". Likewise, the discovery of a chemical element requires not only the production of an elementary substance but also the demonstration that this substance is indeed elementary.[1] In many cases, these two events took place years apart. To complicate things even more, the identity of chemical substances and elements is not necessarily stable over time.[2] Attempts to date a discovery are therefore necessarily arbitrary: the identification of a moment at which the discovery took place depends on the criteria of discovery that are used by a certain community at a given time, and it is impossible to determine who *objectively* deserves credit.

---

[1] For a discussion of element discoveries in general, see [Kragh, 2019].

[2] For example, a recent edited volume by Lissa Roberts and Simon Werrett [2018] shows how the identity of various substances is shaped by both social and material factors. Likewise, in their "ontography" of carbon, Bernadette Bensaude-Vincent and Sacha Loeve [2018] argue against the idea that its nature is fixed and determined by its atomic structure. Instead, they describe carbon's various modes of existence.

Indeed, each discovery takes place in a specific social and historical context and is, therefore, further complicated by questions of priority and prestige. As Simon Schaffer [1986; 1994] argues, the idea of "discovery" only has meaning as an achievement or a prize, a label that research communities retrospectively attribute in order to celebrate certain practices as examples of good research. Schaffer explains how scientific communities idealise and oversimplify long discovery processes by presenting them as clearly identifiable events for which credit can be attributed to a single person. Thus, scientific discoveries are "made up in the course of making the disciplinary histories of specific scientific practices" [Schaffer, 1994, p. 16]. Nevertheless, Schaffer argues that discovery stories should not simply be dismissed as myths; they can reveal something interesting about the social aspects of scientific practice, such as the ways in which authorship and priority are negotiated and how research communities celebrate what they consider to be achievements.

The case of aluminium is a good illustration of the way in which intricate discovery processes are retold and presented as single moments. The isolation of aluminium was the result of a progressive accumulation of knowledge by many different chemists rather than the efforts of a lone genius: it was founded as much on repeated failure as it was on success. Later, when aluminium became an object of scientific and economic prestige, the construction of the narrative of Wöhler and Sainte-Claire Deville as the two fathers of aluminium erased earlier contributions. Until the 20$^{th}$ century, retrospective accounts of aluminium remained centred on the question of priority, and this led historians to reshape the narrative again. Even today, most accounts of the discovery of aluminium focus on the works of Ørsted and Wöhler, and little attention is given to earlier predecessors.

The aim of this paper is to show the different layers of this story. First, I will recount the series of events that led to aluminium's acceptance as an element and production as a material, and second, I will discuss the ways in which this discovery story was told and retold. Throughout this chapter, a social analysis of the attribution of credit for discovery will complement the more traditional, intellectual analysis of the processes that make up the discovery of a phenomenon or object. In order to understand the contradictions that can be found in the literature on aluminium, I argue that we have to take into account not only the scientific advances that led to its production but also the way its story was shaped by various institutional contexts as well as economic, individual and national interests.

## The Long Road to Pure Aluminium

Since the mid-18$^{th}$ century, alumina (aluminium oxide, $Al_2O_3$) had been suspected to be of metallic origin, but the production of a metal from this substance long remained impossible.[3] Failed attempts at doing so are generally excluded from the discovery of aluminium; however, I will demonstrate that it was only thanks to the collaborative development of new methods that the metal could eventually be isolated.

### *Doubts About Alumina's Nature*

In 1754, Marggraf chemically distinguished alumina from lime (calcium oxide) after producing it from a salt called "alum" (aluminium potassium sulfate, $KAl(SO_4)_2$) that could be found in clay and which was used for dyeing. This distinction replaced the ancient element Earth with a series of earthy substances. In today's terms, the earths are (hydrated) metallic oxides and carbonates, but until the early 19$^{th}$ century, they were seen as a distinct class of indecomposable substances, characterised by their incombustibility, insolubility in water, and the fact that they could not be evaporated [Klein and

[3] Throughout this chapter, I will provide modern chemical names for historical substances to improve understanding. However, it should be noted that these are only approximations: there generally is no direct translation between the materials 18$^{th}$ and 19$^{th}$-century chemists were working with and today's chemical substances because methods of purification, separation and identification have changed.

Lefèvre, 2007, Chapter 7.2].[4] Because of its resemblance to salts that had a metallic base, alum was thought to have a similar composition [see, for example, Macquer 1766, pp. 112–117]. Nevertheless, none of the 18th-century attempts at producing a metal from alumina yielded results [Ferrand, 1961, pp. 21–22].

In 1789, the French chemist Antoine-Laurent (de) Lavoisier (1743–1794) listed alumina among the simple bodies, the indecomposable chemical substances that played the role of elements in his new system of chemistry. While he suspected that the earths might one day prove to be metallic oxides, he argued they should be seen as simple until it became possible to decompose them in the laboratory [Lavoisier, 1789, p. 193]. Although he followed his predecessors in expressing the hypothesis that a metal might one day be produced from alumina, he differed from them in his view of the composition of metals. Whereas metals had previously been seen as composed of calces and phlogiston, Lavoisier identified calces as compounds of metals and oxygen (metallic oxides), thus inversing the two classes in terms of composition [Klein and Lefèvre, 2007, Chapter 10.2].[5]

Lavoisier's successors also thought alumina was likely a compound, but they accepted it as simple until proven otherwise [see, for example, Thomson, 1804, p. 473]. Around 1810, the view of alumina as a compound changed from a mere conjecture to a consensus. Moreover, a new metal appeared in textbooks and classifications of simple substances: aluminium [Berzelius, 1811; Thénard, 1813, p. 208; Thomson, 1817, p. 363]. Thus, years before its effective isolation, aluminium was already thought of as an element. As the next section will explain, this shift resulted from Davy's work on the decomposition of the earths.

---

[4] The classification of the earths continually evolved. Around 1800, they were often separated into "alkaline" (magnesia, barites, lime, strontian) and "common" (alumina, yttria, glucina, zirconia, silica) earths, depending on their acidity. During the 19th century, a growing number of "rare" earths were discovered in the minerals gadolinite and cerite. See also Charlotte A. Abney Salomon's contribution in this volume.

[5] In Sweden, metals had already been seen as simple substances [Fors, 2014].

## *The Acceptance of Aluminium as a Metal*

At the Royal Institution in London (Figure 2), Humphry Davy used a voltaic pile — the first battery — to decompose as many compounds as possible in the hopes of identifying "the true elements of bodies" [Davy, 1808b, p. 1]. After showing that potash and soda were oxides of the previously unknown metals potassium and sodium, Davy suspected that the very similar (alkaline and common) earths could likewise be decomposed into oxygen and a metal. Using mercury as the negative electrode of his battery, he was able to form amalgams of the metallic components of the alkaline earths and distil away the mercury. This method had been developed by the Swedish chemists Magnus Martin Pontin (1781–1858) and

Figure 2: The Royal Institution Laboratory, on Albemarle Street in London, where Davy carried out his experiments using the voltaic pile. Engraving by J. Basire, 1818, after W. Tite. Credit: Wellcome Collection.

(Jöns) Jacob Berzelius (1779–1848) and shared with Davy via correspondence.[6] It enabled Davy to isolate enough of the new metals to investigate their properties and confirm their identification as the metallic components of the alkaline earths, which he named barium, magnesium, strontium, and calcium [Davy, 1808a].

The next step was the decomposition of the common earths, one of which was alumina, for which Davy tried multiple methods. Besides forming an amalgam, he fused alumina and potash in a platinum crucible, which he used as the positive electrode of his battery. When he inserted the platinum negative electrode into the mixture, metallic globules rose up to the surface and immediately burned. Once the mixture cooled down, he noticed the negative electrode was covered in a thin metallic film from which alumina could be recovered. However, the quantity of metal produced was too small to study its properties and prove that it was indeed the metallic component of alumina. Even though Davy considered this first attempt a failure, he proposed the name he would have given to the new element if he had been able to isolate it: "alumium" (this later became "aluminium").

Davy [1810] continued his research during the following year. He heated alumina with potassium vapour in a platinum tube and produced tiny metallic particles, which, once again, were too minute to be studied in detail. These particles likely resulted from the decomposition of alumina, but he was unable to prove it as long as they could not be analysed. Similarly, Davy fused iron filings, potassium and alumina into a metallic mass, from which alumina, potash and iron oxide could be produced. This mass was probably an alloy of aluminium, but he was unable to study the metal separately.

Davy complemented these experimental results with arguments based on chemical analogy [Hijmans, 2022]. Since the same methods of decomposition had worked for substances that were chemically very similar to alumina, it was probable that they had been successful in this case as well. Thus, he concluded that alumina most

---

[6]The letter to Davy is now lost, but Berzelius and Pontin [1808; 1810] published their method in Swedish and German.

certainly was a metallic oxide and that it was "extremely likely [that aluminium was a metal], both from the facts detailed, and from analogy" [Davy, 1812, p. 356]. In other words, despite being unable to empirically prove that he had isolated a metal from alumina, Davy thought there were sufficient reasons to believe that aluminium existed. These arguments convinced his colleagues to classify aluminium as one of the elements. In his popular textbook, Louis-Jacques Thénard (1777–1857) characterised aluminium as one of six metals that "have not yet been obtained, and are admitted as metals by analogy, or because the materials from which we extract them have the strongest relations with the metallic oxides" [Thénard, 1813, p. 208].[7]

Though Davy's work established the existence of aluminium, this episode is often excluded from its discovery. Kragh [2019, p. 85], for example, states that aluminium "was known to exist and Davy had even coined a name for it ..., but it had not yet been discovered". I argue, on the contrary, that it constitutes an important part of the discovery process of aluminium since it led to its acceptance as a chemical element.

## *Further Attempts at Isolating Aluminium: Communication and Collaboration*

Davy's publications on aluminium are relatively unique in that they describe failed decomposition attempts in a detailed way. Increasingly throughout the $19^{th}$ century, chemists tended not to publish failed attempts at isolating substances, and this makes it difficult to know how many people tried to isolate aluminium during the 15 years following Davy's work. However, multiple European chemists were studying similar indecomposable substances, such as glucina (beryllium oxide), zirconia (zirconium dioxide), and silica (silicon dioxide), and it is likely they also experimented with alumina. Moreover, there is evidence that a small group of chemists exchanged on alumina and

[7] The other "presumed metals" listed by Thenard were zirconium, yttrium, magnesium, beryllium, and silicon.

related topics, though we can only speculate what they discussed while meeting in person; some of their exchanges can be traced via correspondence.

One of these chemists was Berzelius in Sweden. In April 1824, Berzelius [1912, pp. 69-71] wrote to Davy about his recent experiments in which he used potassium to decompose double fluorides of silicon and potassium or sodium. Once he had isolated silicon using this method, Berzelius tried it on other earthy substances that could form similar double fluoride salts.[8] He succeeded in producing zirconium, but his attempt to isolate aluminium failed because of the difference in solubility of the aluminium salts.[9] Contrary to Davy, Berzelius did not publish a detailed description of his attempt at decomposing alumina. He did not mention this work in his 1824 article on the decomposition of silica, and only vaguely alluded to it in a later publication [Berzelius, 1824; 1827, p. 6]. Nevertheless, he did exchange with other chemists about his work on the earths. Wöhler was his student and assistant when he isolated silicon and zirconium in 1824. The same year, Davy visited Ørsted in Copenhagen, and the two men went to Sweden, where they spent a few days travelling with Berzelius and Wöhler. During this and other encounters that summer, it can be assumed that they discussed their work on the previously indecomposable earths [Christensen, 2013, pp. 425–427].

These exchanges demonstrate that there was a network of European chemists working on the isolation of aluminium and other similar issues. The decomposition of the earths was a well-known chemical problem, and Davy had shown how rewarding such an endeavour could be. For instance, according to Dan Christensen [2013, p. 426], Ørsted's interest in this topic could be explained by his admiration for Davy and the wish to pursue similar

[8] A natural occurrence of such a double fluoride salt is the mineral cryolite, which would later become an important source of aluminium [Kragh, 1995].

[9] Wöhler [1827, p. 158] explained the failure of Berzelius' attempts by the use of too much potassium, which made the solution alkaline and caused the aluminium to dissolve.

achievements, especially since he started investigating the earths in 1823 following a trip to England where he visited Davy. By then, Berzelius had already spent decades investigating this question, in collaboration with multiple of his students and colleagues.[10] The contact between these chemists was crucial for the eventual successful isolation of aluminium, and highlights the collective nature of this discovery.

## Ørsted and Wöhler

In early 1825, Ørsted presented a lump of aluminium at the Danish Academy of Sciences. His method consisted of two steps: first, he produced anhydrous aluminium chloride by heating chlorine, alumina and charcoal in a porcelain tube. The resulting chloride was then heated with a potassium amalgam, producing potassium chloride and an aluminium amalgam.[11] Out of this amalgam, the mercury could be evaporated, leaving behind a lump of tin-like metal. Ørsted did not inform Berzelius or Davy of this result, and he did not announce it as a discovery in any official way. The work was abstracted in the *Journal für Chemie und Physik*, but despite three reminders from its editor, Johann Schweigger (1779–1857), Ørsted did not provide a longer German version. Altogether, his findings failed to attract much interest [Christensen, 2013, pp. 428–429].

In September 1827, Wöhler visited Ørsted, who showed him his method for producing aluminium from its chloride and encouraged him to investigate the metal further since Ørsted did not have the time to do so himself. That same year, Wöhler's first paper on aluminium was published in the *Annalen der Physik* [Wöhler, 1827]. After his trip to Denmark, Wöhler tried to replicate Ørsted's experiments but found that he could not obtain the same results. The distillation of the amalgam produced a grey metallic mass, which

---

[10] Besides Wöhler, Johan August Arfwedson (1792–1841) also studied the earths in Berzelius' lab, for instance. On Berzelius' life and work, see [Jorpes, 1970].

[11] In modern terms, the reactions for these steps may have been $Al_2O_3 + 3C + 3Cl_2 \rightarrow 3\,CO + 2AlCl_3$ and $AlCl_3 + KHg \rightarrow AlHg + KCl + Cl_2$.

released a green gas when heating intensified and left behind only potassium. He therefore started looking for a different method and found that he succeeded by exchanging the potassium amalgam with pure potassium. In a platinum crucible, he heated aluminium chloride and potassium together, which produced a liquid black-grey mass. Once the mixture cooled down, the remaining potassium chloride could be dissolved in water in order to leave behind a metallic powder, which Wöhler believed to be pure aluminium.[12]

As Wöhler himself acknowledged, his isolation of aluminium clearly built on his predecessors' work. Wöhler cited Davy, who, in his view, had reduced aluminium without being able to separate it from the other reaction products. He also cited Ørsted's "ingenious" method for producing aluminium chloride, which constituted an essential step in his new method. However, on the basis of his own replication, Wöhler concluded that Ørsted's method did not yield pure aluminium. Wöhler explained that he had not intended to get involved in someone else's unfinished research but that it was Ørsted himself who encouraged him to continue the work on aluminium. It seems that Ørsted wasn't too interested in obtaining priority for the isolation of aluminium, as he did not contradict Wöhler on the impurity of his sample; in a lecture on the subject in January 1828, he admitted that his lump of aluminium may have indeed contained potassium [Christensen, 2013, p. 430]. There were no signs of bad feelings or rivalry between him and Wöhler. Thus, in the absence of any real debate, the consensus was that Wöhler had been the first to successfully isolate aluminium.

## *Aluminium as a Material*

Wöhler continued his investigation of aluminium, announcing in 1845 that he had produced small globules of the fusible and flexible metal, which could be hammered into thin sheets without tearing [Wöhler, 1845]. Around the middle of the century, the properties of

[12] The reaction may have been $AlCl_3 + K \rightarrow 3KCl + Al$.

Figure 3: Henri Sainte-Claire Deville demonstrating an experiment in the inorganic chemistry laboratory at the *École Normale Supérieure*, Paris, 1878. Reproduction after a painting by L. Lhermitte. Credit: Wellcome Collection.

the earths and different types of metallic compounds remained a topic of investigation for a number of chemists, one of which was Henri Sainte-Claire Deville.

After studying under Jean-Baptiste Dumas (1800–1884), Sainte-Claire Deville's impressive results from his research on mineral waters in Besançon earned him a position at the *Ecole Normale Supérieure* (ENS) in Paris. Upon his arrival there in 1851, he acquired government funding to set up a teaching laboratory (Figure 3), where he investigated the properties of various metals and metallic compounds. Sainte-Claire Deville's arrival in Paris marked the beginning of an especially close friendship with his old professor, which lasted for the rest of his life and likely influenced his research interests. Indeed, his choice to study metals coincided with a reawakening of Dumas' interest in classifying the elements, which required detailed knowledge of chemical elements and their compounds [Klosterman, 1985].[13]

In February 1854, Sainte-Claire Deville presented a new method to produce large quantities of pure aluminium to the French Academy of Sciences. Sainte-Claire Deville's [1854a] method for the isolation of aluminium was similar to Wöhler's, but instead of potassium, he used sodium for the reduction of aluminium chloride.[14] The distillation of the excess aluminium chloride yielded metallic globules, which could be fused together. This method was so effective that it enabled Sainte-Claire Deville to produce a metallic mass big enough to see what it would be like to fabricate aluminium objects. Because of its strength, low density, high malleability, and low tendency to oxidise, he immediately noticed the industrial potential of the new metal. This captured the attention of the Academy: Sainte-Claire Deville was awarded funding for further research, and various members came to visit his laboratory and later reproduced his experiments [Sainte-Claire Deville, 1854c, p. 322; 1854d]. In August 1854, he announced a second method to produce aluminium, using electrolysis as Davy had initially tried. However, instead of mixing alumina and potash, Sainte-Claire Deville [1854c] electrified a heated mixture of sodium chloride and aluminium chloride.[15] Sainte-Claire Deville thus further developed both types of methods for the production of aluminium that had been explored by his predecessors (displacement as well as electrolysis).

Soon, Emperor Napoleon III (1808–1873) became interested in the metal.[16] He arranged for a hall in the state-run Javel works in Paris to be dedicated to the production of aluminium: at this long-established chemical factory, there was more space to carry out

[13] For Sainte-Claire Deville's life and career, see [Paquot, 2005].

[14] The reaction may have been $AlCl_3 + 3Na \rightarrow 3NaCl + Al$.

[15] The salt mixture, which was inert in this reaction, helped ensure that aluminium chloride did not evaporate. The reaction may have been $2AlCl_3 \rightarrow 2Al + 3Cl_2$.

[16] Dumas, by then an important member of the imperial government, helped acquire funding and recognition for Sainte-Claire Deville.

large-scale tests. Less than 18 months after Sainte-Claire Deville's initial presentation, the first bars of aluminium produced at Javel were presented before the Academy [Dumas, 1855]. Besides Javel, Sainte-Claire Deville was involved in the creation of two other factories in and around Paris and one near Rouen. Sainte-Claire Deville's collaborators and assistants moved between the ENS laboratory and the different industrial sites, where research and development took place as well [Le Roux, 2015]. Outside of France, aluminium also sparked an interest, but the early closure of two British factories meant that France remained the only producer of aluminium until the 1880s [Ferrand, 1961, pp. 115–128]. Sainte-Claire Deville and many others continued to perfect the production procedures for industrial quantities of aluminium at a lower cost, but the fascinating story of how aluminium became such an affordable metal is beyond the scope of this chapter.

However, it is important to note that the industrial potential of the new material meant that it was seen as a major achievement, even more so than the discovery of any other chemical element. This becomes especially clear from the reaction of Napoleon III, who exposed the first ingot of aluminium next to his crown jewels at the 1855 *Exposition Universelle* in Paris [Paquot, 2005, Chapter 4; Renaux, 2019]. The recent Emperor was keen to show off his country's technological achievements to the international audience at this world fair, where the new material served as proof of France's modernity. This shows the tremendous amount of prestige that came with the discovery of aluminium, and the high stakes of identifying the legitimate discoverer.

## (Re)Telling the Story of Aluminium

In the years following Wöhler's publication, no one seems to have thought that aluminium had just been "discovered": Wöhler was seen as the first to *isolate* a metal, which had already been predicted to exist. In his *Jahresbericht*, for example, Berzelius [1829, p. 108] reported that aluminium, "which so many chemists sought to isolate in vain", had "at last" been produced by Wöhler, thereby adjusting his earlier reporting of Ørsted's attempts as a success [Berzelius, 1827, p. 118]. Likewise, textbooks descriptions of aluminium were adapted to include Wöhler's isolation of the metal [see, for example, Dumas, 1828, p. 399; Gmelin, 1843, p. 289; Thenard, 1834, p. 200]. Some mentioned only Wöhler, whereas others cited a varying number of predecessors; none, however, identified Wöhler as the "discoverer" of aluminium. As stated in the textbook of Edward Turner (1796–1837), a professor at University College London, Wöhler had merely demonstrated "the propriety of [Davy's] inference" [Turner, 1846, p. 314].

The fairly unproblematic recognition of Wöhler as the first to isolate aluminium became an issue again when aluminium came to be recognised for its industrial potential. I will now show how Sainte-Claire Deville shaped the narrative surrounding aluminium through a number of priority claims and erased the different contributions leading to its isolation — with the exception of Wöhler. Initially, this was very successful, as indicated by the heroic celebration of Sainte-Claire Deville and Wöhler as the fathers of aluminium. Later, as historians charged themselves with the duty of adjudicating in priority disputes, the story of aluminium changed again.

## *Claiming Priority (1): Sainte-Claire Deville and Wöhler*

Two weeks after Sainte-Claire Deville's presentation of aluminium, a letter of a certain Mr. Schratz [1854] was read at the French Academy of Sciences.[17] Schratz, who identified himself as Wöhler's nephew, claimed priority for the discovery of the method for producing aluminium on behalf of his uncle. Sainte-Claire Deville [1854a, p. 279] had cited Wöhler's 1827 publication, but apparently, this did not suffice for Mr. Schratz. The reactions to this letter show that, on the one hand, Wöhler's priority was already well-established, and on the other, Sainte-Claire Deville actively reinforced it.

[17] Besides his last name and address, no information on Mr. Schratz was provided to the Academy.

Dumas [1854] reacted that no one would have thought of doubting "the indisputable rights of Mr. Wöhler" to the title of discoverer. However, he also thought there was a good reason for which Sainte-Claire Deville received much more praise and attention than Wöhler: Sainte-Claire Deville's work had, at last, revealed the "precious qualities" of aluminium, attracting the attention of industry and government. In his reaction, Wöhler [1854] distanced himself from the priority dispute. Not only did he have nothing to do with Schratz's priority claim, he also did not even have a nephew by this name. In his reply, Sainte-Claire Deville [1854b] detailed the way in which his methods and results differed from Wöhler's in order to show the originality of his work. Nevertheless, he insisted on making clear that he was not claiming priority for the discovery of aluminium itself.

In later work, Sainte-Claire Deville [1855] proposed a more detailed comparison between Wöhler's results and his own. When he followed Wöhler's method, he obtained aluminium, which contained sodium impurities. This means that Sainte-Claire Deville found himself in exactly the same situation as Wöhler had in 1827: he had repeated the experiments of his predecessor and shown that they led to impure aluminium. In Wöhler's case, this resulted in discrediting Ørsted's isolation of aluminium. Yet, Sainte-Claire Deville did not take this occasion to claim priority for himself; rather, he insisted that "aluminium is known only thanks to the beautiful works of Mr. Wöhler" [*ibid.*, p. 9]. He never cited any of his other predecessors.

We can only speculate whether Sainte-Claire Deville was motivated by sheer humility or whether he consciously made this diplomatic choice. Either way, the decision to credit Wöhler for the discovery of aluminium worked out in his favour, as Sainte-Claire Deville [1858, p. 504] himself explained:

> Knowing well that I had predecessors in the study of [sodium and aluminium], I did my best to highlight the importance of their work, and for that I have been rewarded. ... Mr. Wöhler has honoured me with a friendship of which I am extremely proud, and

since becoming his collaborator and host, I am happier than ever to have asked only for a small share in the esteem that scientists owe to the discoverer of aluminium.[18]

Indeed, Sainte-Claire Deville's humble decision to reaffirm Wöhler's priority as the true discoverer of aluminium marked the beginning of a friendship between the two men. Sainte-Claire Deville travelled to Göttingen in 1856 and 1858 and became a member of the Göttingen Academy of Sciences [Paquot, 2005, Chapter 12]. Rather than opposing the well-known chemist in a priority dispute, Sainte-Claire Deville aligned himself with Wöhler and was thereby able to share the recognition for aluminium with him.[19]

## *Claiming Priority (2): Sainte-Claire Deville and Others*

Though Sainte-Claire Deville was more than happy to share the credit for the discovery of aluminium with Wöhler, he defended his priority for its industrial production against others. His first priority dispute, with Robert Bunsen (1811–1899), centred on the question of who had been the first to use the electrolytic method for the production of aluminium. Other debates regarding the role that his students and collaborators played in the beginnings of the aluminium industry followed, ultimately motivating him to publish a monograph on aluminium in order to establish his name as the founder of this industry.

In July 1854, Bunsen [1854] published a method to obtain aluminium by electrolysis of aluminium sodium chloride ($AlCl_4Na$) in

[18] Sainte-Claire Deville speaks of multiple predecessors in this passage, which concerns his work on both sodium and aluminium. Besides Wöhler, he cited the Belgian chemists Daniel Joseph Benoît Mareska (1803–1858) and François Marie Louis Donny (1822–1896), whose work provided Sainte-Claire Deville and his colleagues with a method for the production of metallic sodium.

[19] It is unclear whether there were any additional reasons for Sainte-Claire Deville to hold Wöhler in such high regard and whether Dumas played a role in this. Dumas and Wöhler had met in Paris in 1833 and continued to exchange letters, but they were not close friends [Partington, 1964, p. 339].

the *Annalen der Physik*. Ever since he took up his professorship in Heidelberg in 1852, Bunsen had been working on the isolation of a number of metals. He had successfully decomposed magnesium chloride using an electrolytic cell, and he further applied this method to salts of a series of other metals, one of which was aluminium [Ferrand, 1961, pp. 25–31]. When Sainte-Claire Deville [1854c] published his electrolytic method a month later, he was unaware of Bunsen's work on aluminium, but in September, he claimed priority by listing evidence that he had successfully used the electrolytic method months before Bunsen's publication of it. He stated he was proud to have found the same method as such a great scientist and to have found it independently. In his reply, Bunsen's tone was a lot less complimentary. He cited his decomposition of magnesium from 1852 using the same method. Being the inventor of this method, he did not think it was justified that his status was diminished to mere co-applicator of the very same method to a different metal. Looking back on his discovery five years later, Sainte-Claire Deville [1859, p. 4] admitted that Bunsen's "beautiful works" on the decomposition of magnesium had inspired him — but he never cited his work on aluminium.[20]

Two other priority disputes between Sainte-Claire Deville and his students show how unwilling he was to share the priority for industrial aluminium. The first of these controversies involved the Tissier brothers, Charles (1832–1864) and Alexandre (1835–1860), who had worked at the Javel plant on the first trials of the industrial production of aluminium. Whenever Sainte-Claire Deville could not be present, the two brothers were in charge of supervising the experiments.[21] When they left Javel after an argument with the factory director, they opened their own factory and patented some of the machines that were necessary for the industrial process in their own name. According to Sainte-Claire Deville, these machines were based on his designs, patented behind his back and against his will.

---

[20]The controversy between Bunsen and Sainte-Claire Deville is described in more detail by Ferrand [1961, pp. 44–48]. Their disagreement only concerned the electrolytic method: both men viewed Wöhler as the discoverer of aluminium itself.

[21]Sainte-Claire Deville [1858, p. 506] admitted this as well.

In 1858, the Tissier brothers published a monograph on aluminium, which, according to Sainte-Claire Deville, misrepresented the facts. In the preface, they referred to being charged by the Emperor with executing experiments at Javel [Tissier and Tissier, 1858a, p. iii].[22] Furthermore, contradicting Sainte-Claire Deville's account, they claimed that the patented machines were indeed their own work, developed based on ideas of which many were their own, and that Sainte-Claire Deville had initially encouraged them to file for patent in their name [*ibid.*, pp. 72–73]. Sainte-Claire Deville reacted to the publication of this book by sending a letter to the journal *Annales de Chimie*, in which he corrected what he considered to be misinformation. He also heavily criticised the quality of the Tissiers' publication:

> Messieurs Tissier made a point of calling themselves my pupils for having learned from me all the elements of what they know today, as long as this title and my advice were useful to them; but reading their book has proven to me that in many points they did not take advantage of my lessons ...
>
> [Sainte-Claire Deville, 1858, p. 507]

In reaction to the Tissiers' work, Sainte-Claire Deville advanced the publication of his own monograph on aluminium. Despite admitting that it was a bit early to publish a book about the young industry, he justified its publication by the need to assert his priority: "I have decided to take up the pen to talk about my work, only to avoid seeing it diminished and disfigured as it was recently in a book written on aluminium by the Tissier brothers" [Sainte-Claire Deville, 1859, p. 9].[23]

---

[22] The Tissier brothers later explained they had meant this as a collective responsibility: they did not assume anyone would think only the two of them were charged with this mission, especially since the chapter describing the Javel experiments was named after Sainte-Claire Deville [Tissier and Tissier, 1858b].

[23] Despite referring to aluminium as his work ("mon œuvre"), Sainte-Claire Deville again heavily insisted on the priority of Wöhler in his monograph. No other contribution on aluminium was cited. By presenting Wöhler's laboratory as aluminium's

A few years later, Sainte-Claire Deville asserted his priority a third time. In March 1864, Sainte-Claire Deville and his assistant and collaborator Jules Henri Debray (1826–1888) were informed that they would jointly be awarded the gold medal of the *Société d'encouragement pour l'industrie nationale* (Society for the encouragement of national industry). Sainte-Claire Deville immediately sent a letter to Dumas, the president of the society, expressing his surprise and discontent: "The *Société d'Encouragement* gives me a collaborator for aluminium. This is false, or incompletely true. I don't have any when it concerns my scientific works; I have many, when it comes to the industrial production", cited in [Ferrand, 1961, p. 145]. Sainte-Claire Deville's statement regarding his industrial collaborators is rather surprising, seeing that he refused to share his gold medal with one of them.

## *Establishing the Canon on Aluminium*

The strong reactions from Sainte-Claire Deville whenever anyone "diminished" his work contrast with his excessive complimenting of Friedrich Wöhler. It seems that from his first publication, Sainte-Claire Deville encouraged a narrative in which Wöhler and himself were seen as the two fathers of aluminium — and this discovery story caught on. Sainte-Claire Deville and Wöhler were jointly awarded the *Légion d'Honneur* by Napoleon III in 1855 [Paquot, 2005, pp. 43–44]. Thus, for Sainte-Claire Deville, the social implications of the priority for aluminium surpassed his friendship with Wöhler: the shared credit also firmly established his scientific authority and position in society, nationally as well as internationally. In this new narrative, there seems to have been no room for the earlier failed attempts at producing the metal, nor for any collective endeavours that may have led to the birth of the aluminium industry.

As aluminium continued to show promise of usefulness, the fascination with this metal persisted. With this attention came the

---

sole origin, Sainte-Claire Deville not only erased other laboratory experiments but also contributions in terms of work needed to extract and refine bauxite, the mineral ore used for the production of aluminium; see [Roberts, 2016, pp. 342–347].

Figure 4: Bust of Henri Sainte-Claire Deville, moulded in aluminium (1882–1900) © Collection Jean Plateau — Institut pour l'histoire de l'aluminium.

celebration of the two heroes of aluminium, for example, through the production of aluminium objects that commemorated the birth of this industry (see Figure 4). For instance, at the 1900 Paris Exhibition, the aluminium company *Société des Produits Chimiques d'Alais et de Camargue* exhibited a bust of Sainte-Claire Deville and a medal portraying Wöhler [Matignon, 1900]. Likewise, at the 1892 meeting of the American Chemical Society, the Aluminium Company of America sold medals depicting Wöhler [Warren, 1929], and at the 1954 celebration of the centenary of aluminium, every participant was given an aluminium medal representing Sainte-Claire Deville [Ferrand, 1961, p. 106]. The celebration of the discovery of aluminium enabled those who worked in the aluminium industry to share the same cultural heroes. Through the presentation of idealised experimental descriptions that suppress any controversies, "we can see how retrospection aids the production and reproduction of consensus, authority, and conviction inside scientific communities", as Schaffer [1994, p. 21] states. In other words, this is an example of how discovery enables disciplines to build their canon of heroes.[24]

* * *

Throughout the 19$^{th}$ century, Wöhler was seen as the discoverer of aluminium [Lejeal, 1894; Moissonnier, 1903; Richards, 1896]. However, this changed in the 1920s. The celebration of Wöhler had caused Ørsted's work to be mostly ignored, and Danish chemists wanted to celebrate their fellow countryman in honour of the anniversary of his work on electromagnetism. Likely motivated by national pride, chemist and historian J. Fogh reconstructed Ørsted's experiments in 1921 and showed that contrary to what Wöhler had said, pure aluminium could be obtained using this method. Fogh and his colleagues concluded that Ørsted deserved to be given priority as the discoverer of aluminium [Bjerrum, 1926]. They had considerable success and managed to effectively rewrite the history of aluminium, with the result that most 20$^{th}$-century accounts of the discovery of aluminium credited Ørsted rather than Wöhler [Kragh, 2019].

For these historians, it was important to correctly identify which person deserved to be credited for the discovery of aluminium. Harry Holmes [1936, p. 236], for example, found it necessary "for the world to atone for injustice" by re-establishing Ørsted's priority. As discoveries were taken for granted as measures of scientific progress, the historian of science was seen as the right person to adjudicate in the case of a priority dispute. In the case of elements, the most important question in order to make such a judgement was generally who had been the first to successfully isolate an elementary substance. Much historical research has therefore centred on an

---

[24] During the same period, Wöhler played the same role for organic chemists, as he became the hero of their "disciplinary founding myth" [Ramberg, 2000].

evaluation of the attempts of Wöhler and Ørsted, once again excluding other contributions from the history of aluminium.

Some authors have argued that the problem of these evaluations lies with the idea that elements are discovered the moment that they are isolated because — as the case of aluminium illustrates — elements can be known to exist long before. Rancke-Madsen [1976] has, therefore, proposed the idea of an "effective discovery". He argues that the discovery of an element should be attributed to the first person to publicly announce the observation of a previously unknown, chemically distinct substance, which is recognised as containing an element either by the discoverer or at a later point in time. On the basis of these criteria, it was Marggraf who first discovered aluminium in 1754 when he identified alumina [*ibid.*, p. 308]. However, though Rancke-Madsen's view provides an interesting alternative to previous accounts, it does not solve the problem of the oversimplification of the story of aluminium. Moreover, the idea of "effective discovery" relies on fixed criteria of discovery, which do not take into account the specific historical context of each case. As has been argued by Helge Kragh [2019], it is impossible to identify a single set of criteria that are applicable to the discoveries of all elements. Criteria of discovery vary because they depend on the research community that assesses them.

Today, historians and philosophers of science no longer take the notion of discovery for granted. Moreover, many historians have come to consider that it is not their job to judge who deserves priority for scientific discoveries. The realisation that discovery is a complex and collective process has led to the view that there is no objective way to answer the question of priority. As Helge Kragh [2019, p. 86] argues:

> So, when was aluminium discovered and to whom should priority be allocated? ... the question is misguided as it presupposes an answer in terms of a definite year and a definite discoverer. A summary of the discovery process might provide the only appropriate answer ....

The aim of this chapter was to provide such a summary and to show the complexity of the early history of aluminium on two levels. On the one hand, pure aluminium was the result of a development that was geographically and temporally spread out, with each new contribution building on previous works: Sainte-Claire Deville and Wöhler relied on Ørsted's production of aluminium chloride as well as multiple of Davy's and Berzelius' methods for the reduction of the earths. On the other hand, the need to attribute credit for this prestigious achievement has caused the discovery narrative to be constructed and reconstructed over time. Sainte-Claire Deville and his contemporaries identified a clear origin of aluminium in the works of Wöhler and himself, whereas historians later questioned this story based on the idea that Ørsted's earlier efforts might have been successful. I have argued that in order to understand the birth of aluminium, we should take into account all the complexities of this story. Whereas scientists and philosophers of science have long defended a view of discovery as the "eureka" moment of a lone genius, one of the uses of the history of science is to provide a more detailed description of the social, political and institutional contexts as well as the collaborative nature of scientific practice. When it comes to aluminium, this chapter has only treated a small part of its multifaceted history: as an element and material, aluminium connects scientific laboratories, industrial sites, design studies, classrooms, and our daily lives.

## Acknowledgements

The author wishes to thank Hasok Chang, Antoine Duranton, the editors, Annette Lykknes and Brigitte Van Tiggelen, and three anonymous reviewers for their constructive comments, Nathalie Postic of the *Institut pour l'histoire de l'aluminium* for help with Figure 4, and the participants of the SPHERE graduate seminar and the 2020 ESHS conference where earlier versions of this chapter were presented.

## References

Bensaude-Vincent, B. and Loeve, S. (2018). *Carbone: ses vies, ses oeuvres* (Seuil, Paris).

Berzelius, J. J. (1811). Essai sur la nomenclature chimique, *Journal de Phyisque, de Chimie et d'Histoire Naturelle*, 73, pp. 253–286.

Berzelius, J. J. (1824). Untersuchungen über Flussspathsäure und deren merkwürdigsten Verbindungen, *Annalen Der Physik*, 77, pp. 169–230.

Berzelius, J. J. (1827). *Jahresbericht über die Fortschritte der physischen Wissenschaften*, vol. 6 (Heinrich Laupp, Tübingen).

Berzelius, J. J. (1829). *JahresBericht über die Fortschritte der physischen Wissenschaften*, vol. 8 (Heinrich Laupp, Tübingen).

Berzelius, J. J. (1912). *Jac. Berzelius brev. II, Brevväxling mellan Berzelius och Sir Humphry Davy, 1808–1825* (Kungl. Svenska Vetenskapsakademien, Uppsala).

Berzelius, J. J., and Pontin, M. (1808). Försok med Alkaliernas och Jordarternas sönderdelning, *Economiska Annaler,* 6, pp. 110–130.

Berzelius, J. J., and Pontin, M. (1810). Elektrisch-chemische Versuchen über die Zerlegung der Alkalien und der Erden, *Annalen Der Physik*, 36, pp. 247–280.

Bjerrum, N. (1926). Die Entdeckung des Aluminiums, *Zeitschrift für Angewandte Chemie*, 39, pp. 316–317.

Bunsen, R. (1854). Ueber die elektrolytische Gewinnung der Erd und Alkalimetalle, *Annalen Der Physik*, 92, pp. 648–651.

Christensen, D. C. (2013). *Hans Christian Ørsted: Reading Nature's Mind* (Oxford University Press, Oxford).

Davy, H. (1808a). Electro-chemical researches, on the decomposition of the earths, *Philosophical Transactions of the Royal Society*, 98, pp. 333–370.

Davy, H. (1808b). The Bakerian Lecture, on some new Phenomena of chemical Changes produced by Electricity, *Philosophical Transactions of the Royal Society*, 98, pp. 1–45.

Davy, H. (1810). The Bakerian lecture for 1809. On some new electro-chemical researches, *Philosophical Transactions of the Royal Society of London*, 100, 16–74.

Davy, H. (1812). *Elements of Chemical Philosophy* (Johnson, London).

Dumas, J.-B. (1828). *Traité de Chimie Appliquée aux Arts*, vol. 2 (Bechet Jeune, Paris).

Dumas, J.-B. (1854). Observations à l'occasion de la lettre de M. Schratz, *Comptes Rendus de l'Académie Des Sciences*, **38**, pp. 359–360.

Dumas, J.-B. (1855). Présentation d'échantillons d'aluminium au nom de M. Deville, *Comptes Rendus de l'Académie Des Sciences*, **40**, pp. 1296–1297.

Ferrand, L. (1961). *Histoire de la science et des techniques de l'aluminium* (the author, France).

Fors, H. (2014). Elements in the melting pot: Merging chemistry, assaying, and natural history, Ca. 1730–60, *Osiris*, **29**, pp. 230–244.

Gmelin, L. (1843). *Handbuch der Chemie*, vol. 2 (Karl Winter, Heidelberg).

Hijmans, S. (2022), Analogy and composition in early nineteenth-century chemistry: The case of aluminium, *European Journal for Philosophy of Science*, **12**, p. 5.

Holmes, H. N. (1930). The story of aluminium, *Journal of Chemical Education*, **7**, pp. 233–244.

Holmes, H. N. (1936). 50 years of industrial aluminium, *The Scientific Monthly*, **42**, pp. 236–239.

Jagnaux, R. (1891). *Histoire de la chimie* (Baudry et Cie, Paris).

Jorpes, J. E. (1970). *Jac. Berzelius, his life and work* (University of California Press, Berkeley).

Klein, U. and Lefèvre, W. (2007). *Materials in Eighteenth-Century Science: A Historical Ontology* (MIT Press, Cambridge).

Klosterman, L. J. (1985). A research school of chemistry in the nineteenth century: Jean Baptiste Dumas and his research students, *Annals of Science*, **42**, pp. 1–80.

Kragh, H. (1995). From curiosity to industry: The early history of cryolite soda manufacture, *Annals of Science*, **52**, pp. 285–301.

Kragh, H. (2019). Controversial elements: Priority disputes and the discovery of chemical elements, *Substantia*, **3**, pp. 79–90.

Kuhn, T. (1962). Historical structure of scientific discovery, *Science*, **136**, pp. 760–764.

Lavoisier, A.-L. de. (1789). *Traité Elémentaire de Chimie* (Cuchet, Paris).

Lejeal, A. (1894). *L'aluminium, le manganèse, le baryum, le strontium, le calcium et le magnésium* (Baillière et fils, Paris).

Le Roux, M. (2015). From science to industry: The sites of aluminium in France from the nineteenth to the twentieth century, *Ambix*, **62**, pp. 114–137.

Macquer, P.-J. (1766). *Dictionnaire de Chymie* (Lacombe, Paris).

Matignon, C. (1900). Aluminium à l'exposition, *Le Moniteur Scientifique Du Docteur Quesneville*, **14**, pp. 490–502.

Moissonnier, P. (1903). *L'aluminium: Ses propriétés, ses applications* (Gauthier-Villars, Paris).

Paquot, C. (2005). *Henri Sainte Claire Deville: Chimie, recherche et industrie* (Vuibert, Paris).

Partington, J. R. (1964). *A History of Chemistry,* vol. 4 (Macmillan, London).

Ramberg, P. J. (2000). The death of vitalism and the birth of organic chemistry: Wohler's urea synthesis and the disciplinary identity of organic chemistry, *Ambix*, **47**, pp. 170–195.

Rancke-Madsen, E. (1976). The discovery of an element, *Centaurus*, **19**, pp. 299–313.

Renaux, T. (2019). Le dix-neuvième, siècle de l'aluminium français? *Cahiers d'histoire de l'aluminium*, **62**, p. 42–55.

Richards, J. (1896). *Aluminium: Its History, Occurrence, Properties, Metallurgy and Applications, Including its Alloys,* 3rd ed. (Henry Carey Baird, Philadelphia).

Roberts, L. (2016). Exploring global history through the lens of history of chemistry: Materials, identities and governance, *History of Science*, **54**, pp. 335–361.

Roberts, L. and Werrett, S., eds. (2018). *Compound Histories: Materials, Governance, and Production, 1760–1840* (Brill, Leiden).

Sainte-Claire Deville, H. (1854a). De l'aluminium et de ses combinaisons chimiques, *Comptes Rendus de l'Académie Des Sciences*, **38**, pp. 279–281.

Sainte-Claire Deville, H. (1854b). Lettre de M. Deville à M. Dumas, *Comptes Rendus de l'Académie Des Sciences*, **38**, pp. 557–558.

Sainte-Claire Deville, H. (1854c). Note sur deux procédés de préparation de l'aluminium et sur une nouvelle forme du silicium, *Comptes Rendus de l'Académie Des Sciences*, **39**, pp. 321–326.

Sainte-Claire Deville, H. (1854d). Préparation de l'aluminium, *Comptes Rendus de l'Académie Des Sciences*, **39**, p. 535.

Sainte-Claire Deville, H. (1855). Recherches sur les métaux et en particulier sur l'aluminium et sur une nouvelle forme du silicium, *Annales de Chimie et de Physique*, **43**, pp. 5–33.

Sainte-Claire Deville, H. (1858). Lettre de M. Deville, *Annales de Chimie et de Physique*, **53**, pp. 504–507.

Sainte-Claire Deville, H. (1859). *De l'Aluminium, ses propriétés, sa fabrication et ses applications* (Mallet-Bachelier, Paris).
Schaffer, S. (1986). Scientific discoveries and the end of natural philosophy, *Social Studies of Science*, **16**, pp. 387–420.
Schaffer, S. (1994). Making up discovery. In *Dimensions of creativity*, ed. Boden, M. A. (MIT Press, Cambridge), pp. 13–54.
Schratz. (1854). Lettre de M. Schratz, *Comptes Rendus de l'Académie Des Sciences*, **38**, p. 359.
Thénard, L.-J. (1813). *Traité de Chimie Elémentaire, Théorique et Pratique*, 1st ed. (Crochard, Paris).
Thénard, L.-J. (1834). *Traité de Chimie Elémentaire, Théorique et Pratique*, 6th ed. (Crochard, Paris).
Thomson, T. (1804). *A System of Chemistry*, 2nd ed. (Bell & Bradfute, Edinburgh).
Thomson, T. (1817). *A System of Chemistry*, 5th ed. (Baldwin, Cradock & Joy, Edinburgh).
Tissier, C. and Tissier, A. (1858a). *L'aluminium et les métaux alcalins* (Lacroix et Baudry, Paris).
Tissier, C. and Tissier, A. (1858b). Lettre des Frères Tissier, *Annales de Chimie et de Physique*, **54**, pp. 229–231.
Turner, E. (1846). *Elements of Chemistry*, 7th ed. (Thomas Cowperthwait & co, Philadelphia).
Warren, W. H. (1929). The Wohler plaque, *Journal of Chemical Education*, **6**, p. 559.
Weeks, M. E. (1956). *Discovery of the Elements*, 6th ed. (Journal of Chemical Education, Easton).
Wöhler, F. (1827). Über das Aluminium, *Annalen Der Physik*, **11**, pp. 146–161.
Wöhler, F. (1845). Zur Kenntniss des Aluminiums, *Annalen Der Chemie Und Pharmacie*, **53**, pp. 422–426.
Wöhler, F. (1854). Lettre de M. Wöhler à M. Dumas, *Comptes Rendus de l'Académie Des Sciences*, **38**, pp. 555–557.

## Selected Bibliography

Ferrand, L. (1961). *Histoire de la Science et des Techniques de l'Aluminium* (The Author, France).

Kragh, H. (2019). Controversial elements: Priority disputes and the discovery of chemical elements. *Substantia*, **3**, pp. 79–90.

Kuhn, T. (1962). Historical structure of scientific discovery, *Science*, **136**, pp. 760–764.

Paquot, C. (2005). *Henri Sainte Claire Deville: Chimie, Recherche et Industrie* (Vuibert, Paris).

Schaffer, S. (1986). Scientific discoveries and the end of natural philosophy, *Social Studies of Science*, **16**, pp. 387–420.

Schaffer, S. (1994). Making up discovery. In *Dimensions of Creativity*, ed. Boden, M. A. (MIT Press, Cambridge), pp. 13–54.

https://doi.org/10.1142/9789819813858_0005

# A CHEMISTRY OF THE IMPONDERABLE? RADIUM, POLONIUM AND THE DISCOVERY OF ELEMENTS IN THE ERA OF RADIOACTIVITY

Annette Lykknes

*Department of Teacher Education, NTNU-Norwegian University of Science and Technology, N-7491 Trondheim, Norway*
*annette.lykknes@ntnu.no*

*April 1898: Marie Skłodowska Curie (1867–1934), a 30-year-old doctoral student who is researching the new, mysterious phenomenon of radioactivity has just made a remarkable announcement to the scientific community. Her research indicates that two of the minerals she has investigated contain a substance much more "active" or effective in making air a conductor of electricity than their content of uranium could account for. Indeed, the investigations make her propose that the minerals might contain a new element, not before detected in nature.*

The remarkable findings did not stop there: After further investigations, in July and December that same year, Marie and her husband and team-mate Pierre Curie (1859–1906) announced that they had found *two* substances much more "active" than uranium. The first one, not yet confirmed by spectroscopy, they proposed to name "polonium", after Marie's homeland Poland; the second, "radium", after "radio-activity" — the Curies' name for the mysterious rays that had been discovered two years earlier by the physicist Henri Becquerel (1852–1908) [Curie and Curie, 1898; Curie *et al.*, 1898]. The new radioactive elements proposed by the Curies, however, had not even been separated from the mineral fractions in which they were detected, and the main means of identification was not the balance or the spectroscope but their characteristic and unique radioactivity, detected by an electrometer. For this reason, Marie Curie, in her Nobel lecture in 1911, characterised radiochemistry as an "entirely separate kind of chemistry ... which we might well call the chemistry of the imponderable" [Curie, 1911].

What constituted a discovery in the era of radioactivity, when no measurable amounts of the unknown substances existed and when samples were so impure that spectroscopical analysis was difficult too (at least in the beginning)? At what point in the process did the "discovery" of polonium and radium actually take place? When their characteristic radioactivity was detected, when unique spectra became visible or atomic weights were eventually determined, when a salt of the element was separated from the mineral fraction, or when an amount of its elementary form — the metal — had been produced? Was the discovery of a ponderable or imponderable nature?

I will follow the steps in the Curies' investigations into polonium and radium, from detectable radiation to metallic species or mass production, how the actors reconstructed the different parts of the process themselves, and how these achievements were assessed by the scientific community as seen through the Nobel nominations. Through this, I aim to shed light on the particular complexity of discovery processes that took place in the context of radiochemistry, for which no previously accepted criteria for discoveries applied. The understudied discovery history of polonium is a particular case in point; from its initial discovery in 1898, it took 46 years until it was produced in large amounts and 51 years until its atomic weight had made its way into an atomic weight table. The revisiting of the discovery processes is also an opportunity to dispel the myth that the heroine and twice Nobel Prize winner Marie Curie did all the work by herself and demonstrate how connections with industry and even the military-industrial complex became crucial to the endeavour.

## An Unexpected Result

When Marie Curie embarked on a scientific journey to achieve a doctorate in 1897, she opted for a topic that had not yet received much interest in the scientific community. A year earlier, fellow scientist and Parisian Henri Becquerel and the British physicist Silvanus P. Thomson (1851–1916) had independently discovered that uranium salts emitted rays that blackened photographic plates, but scientists did not regard this as a sensational discovery since so many other "rays" were known to exist. What could possibly be so different about the "Becquerel rays", as they were first called? [Malley, 1976, pp. 1–5; Badash, 1979, pp. 10–12]. Some biographers suggest that the neglect of these rays in the scientific community could have been one of the reasons why Marie Curie decided to pursue research on them [Quinn, 1995, p. 143; Pycior, 1996; Boudia, 2001, pp. 47–48], to enable her to pursue new and unexplored research and participate in uncovering unexpected phenomena. She could hardly have imagined how sensational her discoveries would be.

Another probable motivation to take on the new research was Curie's interest in the measurements of the effect of rays on the conductivity of air by using the electrometer [Boudia, 2001, pp. 47–50]. Electroscopes and electrometers had been used to detect and measure electric charges for more than a century, but none of the existing instruments were particularly precise, especially for quantitative measurements. In the 1880s, however, Pierre Curie and his brother

Jacques Curie (1855–1941) invented a special device, a so-called piezo-electric quartz electrometer with an ionisation chamber, where even extremely small amounts of electric charge could be measured. This device came in handy to quantify the charges produced by the uranium rays [Adloff, 2011; Quinn 1995, pp. 110–111]. Marie Curie investigated a large number of substances — according to her autobiographical notes, all known elements were examined — in the form of metals, salts, oxides, and minerals that she obtained from chemists and colleagues at the Museum of Natural History [Curie, 1898; Curie, 1923/1963/2012, p. 45]. She found that all compounds containing uranium or thorium were radioactive; the others were not. Physicist Gerhard Carl Schmidt (1865–1949) at Erlangen had also discovered that thorium was radioactive a few months earlier [Badash, 1966].

Among the minerals Marie Curie tested were pitchblende and chalcolite, known to contain uranium. But these minerals produced much stronger currents than uranium alone did. Pitchblende from three different sources was found to be two, three and four times as active as uranium in her first experiments [Quinn 1995, p. 149]. Interestingly, "artificial chalcolite" did not exert any extraordinary activity. As physicist and science historian Abraham Pais has commented that at this point, by having shown that uranium minerals were more active than uranium or uranium salts in pure form, Curie had already introduced "radioactive properties as a diagnostic for the discovery of new substances" [Pais, 1986, p. 55]. As mentioned earlier, she announced in April 1898 that pitchblende and chalcolite might contain a new element [Curie, 1898, p. 67].

Shortly thereafter, Pierre Curie abandoned his own research to join his wife in her investigations. Together, they presented three months later, in July, the discovery of polonium and the new methods they had used, consciously and explicitly referring to Marie Curie's individual contribution in her original publication from April [Curie and Curie, 1898; Pycior, 1996]. The announcement that they had detected a second new element was made in December and co-authored by Gustave Bémont (1857–1937), one of the laboratory chiefs at the *Ecole municipale de physique et de chimie industrielles*

(EPCI) where Pierre was working and who joined them temporarily in their experimental investigations [Curie *et al.*, 1898; Curie, 1923/1963/2012, p. 50]. The next obvious step in the investigation was to separate the elements responsible for the activity detected in pitchblende and chalcolite. From the very beginning, the Curie team realised that this would be both a time-consuming and meticulous undertaking, which would require the involvement of (more) chemists and even liaisons with industry. As a consequence, more collaborators were hired, among them chemist André Debierne (1874–1949), who, from July 1899, was put in charge of adapting laboratory techniques and doing try-outs with industry [Boudia, 2001, p. 72]. There were also assistants such as a certain M. Petit,[1] who Pierre Curie apparently brought with him from the Sorbonne [Reid, 1974, p. 91] (see Figure 1). Pierre Curie participated in the separation work for a few years but decided in 1901 to leave this line

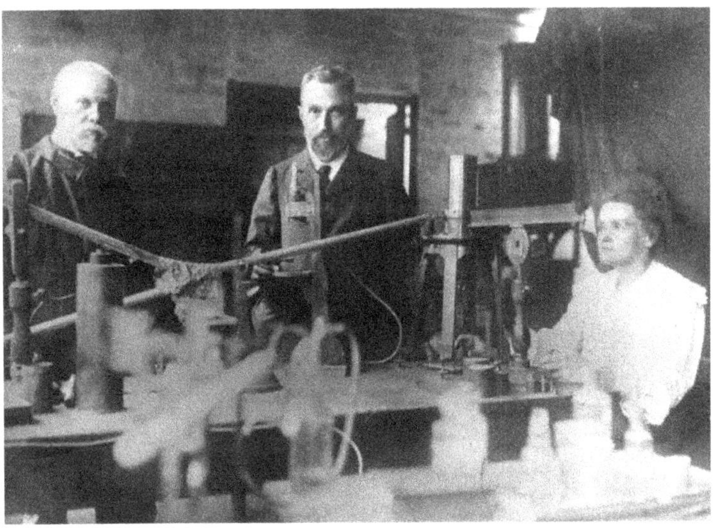

Figure 1: Marie and Pierre Curie in their laboratory with their assistant, M. Petit. Musée Curie, coll. ACJC; reproduced by kind permission of the Musée Curie.

[1] In some sources, the man in the picture with the Curie couple is said to be their chemist colleague, Gustave Bémont.

of work to join Marie Curie and put his efforts into investigating the properties of radium instead [Pycior, 1996].

## First Attempts at Isolation of the New Elements

In their first attempt at the isolation of the elements, the Curies and their team pulverised 100 grams of pitchblende and treated the sample with chemicals to be able to separate substances that reacted differently chemically into separate fractions. Finally, the activity of the substances in the different fractions was measured, and as they further separated the products, they became increasingly active — from 17 times the activity of uranium to 150 times, and then to 300 [Quinn, 1995, p. 151]. This element reacted chemically along with bismuth so that it was always found in the bismuth fraction, due to similar chemical properties (Figure 2). In July 1898, the Curie team produced a bismuth fraction, which carried an activity that was 400 times as intense as that of uranium [Curie and Curie, 1898].

**Tabelle II.**

| Reihen | Gruppe I. — $R^2O$ | Gruppe II. — $RO$ | Gruppe III. — $R^2O^3$ | Gruppe IV. $RH^4$ $RO^2$ | Gruppe V. $RH^3$ $R^2O^5$ | Gruppe VI. $RH^2$ $RO^3$ | Gruppe VII. $RH$ $R^2O^7$ | Gruppe VIII. — $RO^4$ |
|---|---|---|---|---|---|---|---|---|
| 1 | H=1 | | | | | | | |
| 2 | Li=7 | Be=9,4 | B=11 | C=12 | N=14 | O=16 | F=19 | |
| 3 | Na=23 | Mg=24 | Al=27,3 | Si=28 | P=31 | S=32 | Cl=35,5 | |
| 4 | K=39 | Ca=40 | —=44 | Ti=48 | V=51 | Cr=52 | Mn=55 | Fe=56, Co=59, Ni=59, Cu=63 |
| 5 | (Cu=63) | Zn=65 | —=68 | —=72 | As=75 | Se=78 | Br=80 | |
| 6 | Rb=85 | Sr=87 | ?Yt=88 | Zr=90 | Nb=94 | Mo=96 | —=100 | Ru=104, Rh=104, Pd=106, Ag=108 |
| 7 | (Ag=108) | Cd=112 | In=113 | Sn=118 | Sb=122 | Te=125 | J=127 | |
| 8 | Cs=133 | Ba=137 | ?Di=138 | ?Ce=140 | — | — | — | — — — — |
| 9 | (—) | — | — | — | — | — | — | |
| 10 | — | — | ?Er=178 | ?La=180 | Ta=182 | W=184 | — | Os=195, Ir=197, Pt=198, Au=199 |
| 11 | (Au=199) | Hg=200 | Tl=204 | Pb=207 | Bi=208 | — | — | |
| 12 | — | — | — | Th=231 | — | U=240 | — | — — — — |

Figure 2: The 1871 version of the periodic system according to the Russian chemist Dmitrii Mendeleev (1834–1907). The lines represent yet undiscovered elements, among them polonium, which shared chemical properties with bismuth (here at the bottom of group V) and radium, which was similar to barium (group II). In the public domain.

Having failed to produce spectroscopic evidence earlier, they had engaged spectroscopy authority Eugène Demarçay (1852–1904), who, less than two years previously, had identified unique spectral lines of the element that would be known as europium (atomic number 63 in modern periodic systems). Indeed, spectroscopy had become a versatile tool even in chemistry after its invention by Robert Bunsen (1811–1899) and Gustav Kirchhoff (1824–1887) in 1859 — they made subsequent discoveries of elements caesium and rubidium in mineral waters the following year and additional identifications in the 1860s and 1870s (thallium, indium, helium, gallium). However, Demarçay was not able to distinguish any new characteristic lines in the Curies's bismuth sample except for impurities [Curie and Curie, 1898; Demarçay, 1898]. Apparently, the sample of polonium that the Curies and their collaborators analysed was only a few nanograms and, therefore, below the sensitivity limits of spectroscopic detection at the time [Adloff and Kauffman, 2007]. They nevertheless proposed its existence [Curie and Curie, 1898]. Indeed, in the laboratory notebook from 13 July, five days before the publication where the announcement was made, we can see the symbol "Po" [polonium] in one of the pages in Pierre Curie's handwriting.[2]

After a summer break, the Curies resumed their experiments with Bémont, and after several months of hard work, as mentioned in December 1898, the three announced to the French Academy of Sciences that they had found the second radioactive substance in pitchblende, which reacted chemically with barium (Figure 2), and whose fraction was 900 times more radioactive than uranium. In the case of the second element in the barium fraction, however, Demarçay managed to identify one unique spectral line, which they ascribed the new element they named "radium" [Curie *et al.*, 1898]. Naturally, the spectroscopical evidence added weight to their experimental findings.

---

[2] BnF [see the References section]: XVI Carnet de la découverte 2. "Uranium II". 18 mars to 16 juillet 1898. The mentioned page is from 13 July and is also displayed in [Adloff and Kauffman, 2007, p. 95].

Although the Curies had no doubt as to whether the two new elements existed, "to make chemists admit their existence", as Marie Curie [1923/1963/2012, p. 47] expressed, they considered it necessary not only to isolate the elements but also to determine their unique atomic weights, thus moving from the realm of the imponderable to the realm of the ponderable. After some initial experiments indicated that it would be easier to separate radium than polonium, the team decided to concentrate their efforts on the former [Curie, 1923/1963/2012, p. 48]. Probably, the spectroscopical evidence, which was conclusive for radium but inconclusive for polonium, was also decisive for their decision. At the beginning of 1899, Marie Curie had obtained 1–2 centigrams of radioactive radium chloride but realised she needed more to be able to determine the atomic weight, especially since the amounts of radium diminished during the chemical operations [Boudia, 2001, p. 71].

## Finding Space — and Means — to Isolate Radium

Being determined and prepared to isolate radium was one thing, but having the opportunity to do so was another. Marie Curie recalled that "it was during this period of our research that we were extremely handicapped by inadequate conditions, by the lack of a proper place to work in, by the lack of money and of personnel" [Curie, 1923/1963/2012, p. 47]. At Pierre's workplace, the EPCI, the Curies were given permission to use an in-glassed study on the ground floor, which was normally used for storage. There were no hoods to carry away poisonous gas, so they often had to carry out the work outside. As for the radioactive material, thanks to the influence of the Austrian Academy of Sciences, the Curies were able to purchase several tons of pitchblende residues from the St. Joachimsthal mine in Bohemia, which had been used to extract uranium for use in fluorescent paint. Private funds, as well as financial donations, made the purchase possible, which was at an "advantageous price" [Curie, 1923/1963/2012, p. 47; Quinn, 1995, p. 155]. The team was ready to start the work in the spring of 1899.

How were the chemical analyses of the mineral residues carried out? This part of the history of the discovery of radium has been vaguely described, even obscured. In one part of her biography of Pierre Curie, Marie Curie explained that the two worked "practically unaided during two years" with the chemical treatments of the pitchblende residues [Curie, 1923/1963/2012, p. 48], while she, a few pages later, emphasised the importance of collaborators, and that Gustave Bémont was working with them already from 1898 [Curie, 1923/1963/2012, p. 50]. In the subsequent chapter, Marie Curie explained that they had concluded that they "could not make further progress without the aid of industrial means of treating raw material" and that the first industrial experiment was organised by Pierre Curie in 1899 [Curie, 1923/1963/2012, p. 52]. Nevertheless, as the historian of science Xavier Roqué [1997, p. 268] has argued, "historians and biographers have not shown much understanding for this approach to radioactivity". Indeed, the industrial connection has been toned down in the biographies of Marie Curie (among them the authoritative and widely used book by Susan Quinn [1995]), which might explain how the image of Marie Curie working alone with her tons of minerals — as we can see in the background of Figure 3 — has become an established myth in popular accounts about the history of radioactivity.

Although Bémont's contribution to the discovery of radium was acknowledged in the joint publication about the discovery from 1898 [Curie *et al.*, 1898], it is possible that the extent of his contributions has not been fully recognised. He once explained to a journalist that he did the initial chemical analyses of a 5-kilogram sample of the mineral and separated it into chemical groups or fractions, although according to the historian of science Soraya Boudia [2001, p. 71], he did not concentrate the radioactive substances. Likewise, with the arrival of the large amounts of pitchblende, the Curies became dependent on services from collaborators and on expertise in and outside of their laboratory. André Debierne, who arrived in 1899, initiated collaboration with the *Société centrale des produits chimiques*, which was also marketing Pierre Curie's scientific instruments,

Figure 3: A pile of stone displayed in the Nobel Prize Museum (2011–2012) as part of the exhibition celebrating the centenary of Marie Curie's second Nobel prize. In the background, we see a famous picture depicting the heroine stirring in one of her boilers all by herself. (Photo: Annette Lykknes. Reproduced by kind permission of the Nobel Prize Museum.)

to try out semi-industrial treatment of the minerals. From mid-1899 till the end of 1903, in exchange for a proportion of the extracted radium salts, the *Société* supplied chemicals and paid the staff's wages [Boudia, 1997; Boudia, 2001, p. 72].

In 1902, from a little more than one decigram of radium chloride extracted from several tons of pitchblende, Marie Curie was able to make the first accurate determination of the atomic weight of radium.[3] The radium-bearing barium fraction had been extracted at

[3] BnF [see the References section]: XVII Carnet de la découverte 3. "Uranium III — Polonium". 11 novembre 1898 to 13 juillet 1900. The journal spans longer than

the factory, and she carried out the fractional crystallisation in the laboratory to obtain the decigram of radium chloride [Curie, 1923/1963/2012, p. 54]. Curie worked with 20 kilograms of the gross mineral fraction at a time and noted how "exhausting work [it was] to move the containers about, to transfer the liquids, and to stir for hours at a time" [Curie, 1923/1963/2012, p. 48]. This first collaboration with the industry was a success and a good starting point for future collaboration with the upcoming radium industry. In 1907, Marie Curie was able to make a new and more accurate atomic weight determination based on four decigrams that had been produced at the radium factory *Sels de radium* (Radium salts), set up in 1904 in Nogent-sur-Marne outside Paris by the industrial chemist Emile Armet de Lisle (1853–1928), as an extension of the family's quinine business [Curie, 1911; Roqué, 1997; Boudia, 2001, pp. 92–96]. The factory drew on advice not only from the Curies but also from their main collaborators, chemist Frédéric Haudepin from the *Société*, Jacques Danne (1882–1919), who was Pierre Curie's assistant, and Georges Gabriel Robillard and Paul Razet, who were working in the Curie laboratory [Boudia, 1997]. The Curies later also established collaboration with other radium producers.

Roqué [1997, p. 275] has described the collaboration with de Lisle's industry as a "key asset" to Marie Curie; through this relationship, Curie had access to Armet's workers and his installations while only paying for the time used, a success formula that was later repeated by another chemist couple looking for new chemical elements — Ida (1896–1978) and Walter Noddack (1893–1960) [Van Tiggelen, 2001].[4] Indeed, it has been argued that Curie's industrial activities were central to her work, not just a by-product or

13 July. An approximate atomic weight determination, based on very little radium chloride, is reported on 27 July 1900 (p. 65 in the laboratory notebook). The determination I refer to above is reported on 28 March 1902 (pp. 66–67 in the laboratory notebook) and amounted to 225.93. I acknowledge Robert Reid who drew my attention to this part of the notebook [Reid, 1974, pp. 106–107]. Successive atomic weight determinations varied from 138 to 226.45, the last one conducted in 1907 [Curie, 1911].

[4] See also Brigitte Van Tiggelen's contribution to this volume.

side-effect of the research process that Curie envisioned radioactivity research with industrial value, and that her laboratory was important for the construction of the radioelement industry [Roqué, 1997; 2019; Boudia, 1997; 2001]. Also, contrary to the ubiquitous statement that the Curies were never interested in patenting, a recent scholarship has emphasised that despite their lack of interest in commercial success, the Curie couple, Marie Curie in particular, used both publishing and patenting to their advantage (Pierre Curie patented instruments), helped build bridges between "pure" and "applied" science, and used industry contracts to consolidate their radium legacy [Hemmungs Wirtén, 2015a, p. 29].

## Radium Metal!

After almost four years of hard labour since the announcement of the discoveries of polonium and radium and thanks to a fruitful collaboration in both her laboratory and with industry, Marie Curie had finally reached her goal: the first decigram of radium salt had been prepared, its atomic weight determined, and its spectrum produced. "Thus," Curie wrote in her biography of Pierre in 1923, "the chemical individuality of radium was completely established" and added that there could be no further controversy about the existence of the element radium [Curie, [1923/1963/2012] 2012, p. 54]. However, this was Marie Curie speaking in retrospect. Already, in 1906, four years after the first accurate determination of radium's atomic weight had been made, the then 82-year-old British physicist Lord Kelvin (William Thomson, 1824–1907) had attacked the idea of radium as an element in the correspondence columns of *The Times*, arguing that what they had found was rather a molecular compound, perhaps of lead and helium [Reid, 1974, pp. 161–162]. It was, therefore, important for Marie Curie to continue working to establish that radium was an element. One way was to produce it in its purest form, as a metal, which she set out to do with her co-worker Debierne. This was a tedious process, which first required that she separate larger quantities of radium chloride and then use it to prepare a radium amalgam with mercury through an electrolytic process, and, lastly, to distill the minute amount they had and condense it. Finally, a small but recognisable amount of a white but volatile metal, which strongly adhered to iron, turned black upon exposure to air, burnt paper, and oxidised in water — properties "exactly [as] the ones that can be forecast on the assumption that the radioactivity of the salts is an atomic property of the radium which is unaffected by the state of combination", as Marie Curie stated in her Nobel lecture in 1911 [Curie and Debierne, 1910; Curie, 1911; Reid, 1974, pp. 164–165]. There, she also emphasised how important it was to bring forth this evidence since there had been voices "to whom the atomic hypothesis of radioactivity was still not evident" [Curie, 1911], not only referring to Lord Kelvin but to a general scepticism, especially widespread amongst chemists [Malley, 2011, pp. 64–66]. Mendeleev, for example, tried to account for radium's activity in terms of the ether [Mendeléeff, 1904, pp. 45–49].

Marie Curie's 1974 biographer, Robert Reid, describes her work to produce radium metal as an act of utmost importance to her, something she took on with an obsession and resolution equal to her first determination to isolate radium. That Curie and Debierne had obtained metallic radium reached *The New York Times* on 6 September 1910 and British newspapers the day after [New York Times, 1910; BNA, 2013]; however, this milestone in the work on radium and radioactivity has received surprisingly little attention in biographies and popular accounts compared to the first physical evidence of the existence of radium; despite that, the two Nobel nominations put forward for her candidature in 1911 had, in fact, emphasised this very achievement.[5] The author of the most comprehensive biography of Marie Curie, Quinn, for example, lists the isolation of metallic radium among Curie's achievements since the

[5] Nobel Archives, The Royal Swedish Academy of Sciences, Stockholm, Minutes of the Nobel proceedings of the Royal Academy of Sciences 1911, nominations by G. Darboux, dated 25 September 1910 and by S. Arrhenius, both registered on 28 January 1911.

1903 Nobel Prize [Quinn, 1995, p. 308] but does not discuss its implications or importance. Similarly, the difficult work on polonium has been eclipsed by the heroic story of radium, as we shall now see.

## Polonium in the Haystack

While the discovery of radium has been told and retold over and over again, from the initial announcement to the spectral evidence and the first isolation, it is much harder to find descriptions of the different steps of the discovery of polonium in literature on Marie Curie or the history of radioactivity. In popular accounts, the story of the discoveries of the two new elements — and of radioactivity itself — often ends with the determination of the atomic weight of radium, e.g., [Des Jardins, 2011; Parsons, n.d.]. Since radium was the first radioelement whose physical existence was confirmed, its ground-breaking status left no doubt that similar evidence would come for other elements as well, but in the case of polonium, the problem was that it existed in very small quantities in nature. In fact, it is today regarded as one of the rarest naturally occurring elements, with an abundance of only 0.2 per cent of that of radium [Bagnall, 1963]. As the physicist and 1908 Nobel Prize in Chemistry awardee Ernest Rutherford (1871–1937) explained in a publication in *Nature* in 1910: Since polonium decays 5,000 times faster than radium, "the maximum amount of polonium to be obtained from a mineral is in consequence only 1/5,000 of the amount of radium" [Rutherford, 1910, p. 491]. With a starting material of one ton of pitchblende containing 50 per cent of uranium, for example, the amount of radium present at the time would be 170 milligrams while that of polonium is only 1/30 milligram (ca. 0.033 mg). That polonium had been at all detected, with the limited experimental techniques available at the time, has been described as a "remarkable feat" [Bagnall, 1963, p. 61].

Since the Curies expected that it was more likely to find radium than polonium in the haystack, efforts, as mentioned earlier, were set on finding the former. But this did not mean the couple stopped thinking about polonium. Meanwhile, other scientists were interested in the fascinating substance; already in 1899, Friedrich Giesel (1852–1927), inorganic chemist and director of *Buchler und Compagnie* manufacturers of quinine in Braunschweig, prepared his own samples of radium and polonium and started to investigate their properties [Romer, 1970, pp. 76–77; Giesel, 1899]. Like the Curies, he reported that the polonium preparations lost their activity slowly over time. This was a few years before Rutherford and Frederick Soddy (1877–1956) proposed the theory of radioactive decay in 1902, and a similar "disappearance" had not been observed in radium. In fact, the difference in the behaviour of polonium compared to other radioactive substances made the Curies in 1902 express some doubt as to whether it was, after all, an element: "This substance is a kind of bismuth; it has not yet been proven that it contains a new element," they stated [Curie and Curie, 1902, p. 86, translated by Adloff and Kauffman, 2007], and it has been argued that this statement made the Curies open to competition. As biographer Reid expressed, "She [Marie Curie] had embarrassingly put her doubts into print" [Reid, 1974, p. 163].

In June that same year (1902), Willy Marckwald (1864–1942) of the University of Berlin reported on investigations into a bismuth fraction of a residue left after the decomposition of pitchblende left to him as a consultant for a company in Hamburg. He followed the radioactive bismuth fraction and noticed no diminution of the radioactivity over the course of time, unlike what the Curies had found for their substance polonium. Marckwald, therefore, implied that what he had found was not the same substance that had been detected by the Curie couple in 1898 [Romer, 1970, pp. 76–77, 80–84; Marckwald, 1902–03]. Curie responded shortly after that this element appeared to be the same as her "polonium" [Curie, 1902–03]. Marckwald, however, maintained his view since his element remained radioactive over time and reacted chemically more like tellurium than bismuth, and christened it "radiotellurium" [Romer, 1970, pp. 87–92; Fontani *et al.*, 2015, pp. 452–453].

Retrospectively, that the experiments with Curie's polonium and Marckwald's radiotellurium differed is perhaps not surprising since the materials the scientists were working with changed from one day to the next, as the science historian Alfred Romer [1970, pp. 76–77] reminded us. It is also plausible that Marckwald insisted on his conclusions because he did not want to give up on his priority for the discovery of an element so easily [Quinn, 1995, p. 173]. The theory of radioactive decay introduced by Rutherford and Soddy in 1902, the same year as the dispute over polonium took place, added some more understanding to the disappearance of substances. Less than two years later, in March 1904, Curie got support from the highest authority: In March 1904, Soddy himself claimed in a commentary in *Nature* that Markwald's conclusion that radiotellurium did not decay over time was "unsupported by actual measurements", and he therefore rejected evidence supporting that "radio-tellurium possesses a more constant radio-activity than polonium" [Soddy, 1904, pp. 461–462]. What eventually made Markwald surrender (and accept the name "polonium" for the element) was the coinciding half-lives determined independently by Markwald and Curie in 1905 and 1906, respectively [Adloff and Kauffman, 2007]. As chemist historians Jean-Pierre Adloff and George B. Kauffman [2007] noted, from that time on, half-life could be added to the list of identifiers for radioelements, along with their particular radioactivity, spectra, and atomic weights.

The quarrel with Marckwald in 1902 onwards had made Curie more determined to isolate and identify polonium as an element, to set all doubt aside. Over the years, she made several attempts to concentrate polonium, which led to fractions with high activity but without arriving at definite results. Indeed, for decades, polonium was only available in extremely small amounts, with one exception: a quantity of 0.1 mg, which Curie, Debierne and their co-workers managed to prepare in 1910 from tons of residues from uranium mineral that had been first treated in the factory of Armet de Lisle [Curie and Debierne, 1910]. From 200 grams of material 5,500 times more active than uranium, they pursued the concentration in the laboratory and eventually separated an active substance containing 1–2 per cent polonium, which was 50 times more active than an equal weight of radium [Curie and Debierne, 2010; Curie, 2011; Adloff and Kauffman, 2007]. The analytical procedure was extremely difficult, but the small amount of polonium present sufficed: Spectral analysis yielded seven new rays, some of which "appear attributable to polonium", Marie Curie stated in her Nobel lecture in 1911. Adloff and Kauffman, reviewing the attempts to prove the existence of polonium almost 100 years later, conclude more strongly about this evidence: They call this the definite proof of the existence of polonium 12 years after its initial discovery based on its radioactivity [Adloff and Kauffman, 2007]. Bagnall [1963] and Moyer [1956] also emphasise the successful efforts of Curie and Debierne in 1910 as an exception in the early years of investigation in their retrospective accounts on the production of polonium. In 1938, 28 years after Curie and Debierne's achievement, X-ray spectroscopic evidence was provided from trace amounts of the element by a Parisian team who was able to confirm the atomic number (84) of polonium with a newly devised X-ray spectrograph [Hulubei *et al.*, 1938; Van Tiggelen, 2019]. These early results were confirmed in the 1950s [Bagnall, 1962, p. 208].

## Handling and Producing Polonium

Despite the initial difficulties, a powerful supply of polonium was gradually built up at Marie Curie's Radium Institute in Paris. Building on the work and materials in her mother's laboratory, Irène Curie (1897–1956) became instrumental in this. She acquired expertise in isolating polonium and trained others, such as the Hungarian radiochemist Elizabeth Róna (1890–1981), in handling polonium sources. The preparation of polonium sources involved not only the separation of polonium from a mixture of elements, "radioactively pure polonium" also needed to be precipitated on a metallic surface, which added to the tedious and meticulous concentration process [Rentetzi, 2019, p. 337]. Indeed, Irène Curie's comprehensive review of the extraction and purification of polonium (and other radioelements) from 1925 [Curie, 1925] has been cited in reviews on polonium and its chemistry decades later [e.g., Bagnall, 1957; Sethy *et al.*, 2015]. Her doctoral dissertation, which was also submitted in 1925, was about the alpha rays emitted from polonium [Perrin, 1981; Bonolis, n.d.] In fact, Irène and her husband, Frédéric Joliot-Curie (1900–1958), since 1926, used alpha particles of radioactive polonium as projectiles to bombard beryllium nuclei, which led to their Nobel winning discovery of artificial radioactivity in 1934 [Jacquemond, 2019]. For this purpose, the short half-life of the polonium nuclide came in handy.

The first large amounts of polonium were produced only in March 1944 in connection with work in secrecy for the Manhattan Project to build the first atomic bombs [Hochwalt and Haring, 1948; Fernelius, 1982; Sopka and Sopka, 2010]. A research site was set up in 1943 in an old building in Dayton, Ohio, with the intention to produce polonium for a polonium-beryllium neutron source to ignite an implosion bomb (the one later dropped on Nagasaki in August 1945). In the same way as the Curies collaborated with industry to concentrate radium in the early years of the $20^{th}$ century, the industry also became involved in the isolation and preparation of polonium for the Manhattan Project: The Monsanto Chemical Company was given prime responsibility for the purification and production of polonium in the necessary quantities and worked in close cooperation with radiochemists located at the top-secret laboratory at Los Alamos, known as "Project Y" [Hochwalt and Haring, 1948]. From November 1943, polonium was extracted "wet-chemically" from natural radioactive lead residues, but from June 1945, the chemical method was replaced by neutron-bombardment of bismuth to produce polonium.

As for the atomic weight of polonium, the value of 210 entered the international atomic weight table issued by the International Union of Pure and Applied Chemistry (IUPAC) only in 1949 [Wichers, 1950]. Indeed, the natural occurrence of the element polonium is limited to trace amounts of the short-lived mass-210 polonium isotope (half-life of 138 days).

## Discussion

"Marie and Pierre Curie discovered the elements polonium and radium in 1898." Variations of this sentence are easy to find in textbooks and popular accounts in the early history of radioactivity or the history of the periodic system. However, much is missing from such a statement and likewise from accounts that hold on to a simplified discovery history, placing emphasis on heroes, particular times and places but not processes. As the historian, philosopher and science educator Douglas Allchin [2003] has commented, typical ingredients of mythical narratives that are often found when the history of science is used anecdotally in science teaching are heroisation of historical actors, idealisation of scientific processes, creation of linear timelines of what was originally a web of history, and a sharpening of actors' roles for the sake of telling a good story. Many of these aspects can be found in discovery histories about polonium and radium. I suggest that the particular context of radioactivity research makes the history of the discovery process even more complex than is often the case with element discoveries.

Firstly, popular accounts of the discoveries of polonium and radium often appear as heroic tales. As Marie Curie is the most famous female scientist in history and an important role model for women in science, it is not surprising that stories about Marie Curie's research are surrounded by mystique and myth. In an article in *Science* in 1974, historian of science Stephen Brush warned teachers against using the history of science in their classroom teaching because it would complicate matters [Brush, 1974]. Indeed, it was his experience that teachers wanted to have their heroes of science and that history would take them down from their pedestals. The idea of heroising scientists for the purpose of telling a good story does not only apply to the school context, of course. When telling the general public the story about an exceptional woman scientist who had received two Nobel Prizes in Science, it might be tempting to emphasise her great achievements and even narrow the story down to "clean" linear stories, even if the story also includes her partner in life and science, Pierre Curie. However, the account of the events of the discovery processes of polonium and radium described in this article demonstrates that neither the heroine Marie nor the associated hero Pierre sufficed to accomplish what was needed to convince the scientific community about their discoveries. Indeed, a number of collaborators and qualified expertise outside of their own laboratory (for example, within spectroscopy) was crucial to the endeavour, not to mention the essential connection with industry, which has indeed been an undercommunicated aspect of the early history of radioactivity [Roque, 1997]. The polonium discovery even extended beyond Marie Curie's lifetime. Eva Hemmungs Wirtén, looking at how the public Marie Curie came to be from the perspective of intellectual property, calls for a broader understanding of Curie than the traditional heroine on the pedestal, including her "strategies for self-fashioning, institution-building, networking" and her way of building alliances, attracting others, and protecting her investments [Hemmungs Wirtén, 2015a; Hemmungs Wirtén, 2015b, p. 607]. More nuanced accounts on the discovery processes, as I have attempted in this article, contribute to understanding, at the very least, parts of what Hemmungs Wirtén is asking for: How central alliances with industry were mutually beneficial to Curie and the industry — indeed, the isolation and determination of the atomic weight of radium would not have been possible without the industrial connection. Other central points to the story of the discovery of polonium and radium are that additional expertise needed to be sought to accomplish results and, therefore, that the "heroine doing hard labour alone", as shown in Figure 3, is a deceptive image of the early history of radioactivity overall and the discovery of the elements polonium and radium in particular.

Secondly, the discovery process is cut short in many historic tales on polonium and radium since the two new radioelements are normally pinpointed to 1898 when Marie and Pierre Curie (with the help of Gustave Bémont) detected them by means of their particular radioactivity. The hard work involved in — and significance of — getting the first decigram of radium salt is often under-communicated, not to forget in the production of metallic radium. While the first Nobel Prize that Marie and Pierre Curie shared with Henri Becquerel in Physics in 1903 was given in recognition of the research on the radioactivity phenomena [Nobelprize, 1903], the 1911 Nobel Prize in Chemistry was awarded to Marie Curie for her "discovery of the elements radium and polonium, ... the isolation of radium and the study of the nature of the compounds of this remarkable elements" [Nobelprize, 1911]. In fact, the Swedish professor of chemistry Svante Arrhenius (1859–1927) emphasised in his nomination letter that the discovery of radium was much greater than any other element discovery since a whole science was based upon it. He also highlighted the isolation of metallic radium in 1910, which, in his view, had proven that "radium is a real metal and not a composition such as, for example, ammonium". The second nominator, Parisian professor of mathematics Gaston Darboux (1842–1917), gave only one reason for proposing Marie Curie for the prize: "She just prepared metallic radium".[6] This shows that toning down the significance of isolating radium in metallic form is not doing justice to the achievement nor to the way it was assessed at the time.

Thirdly, the contrasting fate of radium and polonium adds to the nuances of the discovery history; it is indeed two histories rather than one history. While the radium story itself includes many steps and spans more than a decade, with radioactivity detection, characterisation and separation, spectroscopical investigations, atomic weight determination, and isolation of a sample of metallic radium, the story of polonium far exceeds the former in duration and complexity. For this reason, and probably because the end to the polonium discovery history is not easily defined, it is rarely included in histories of early radioactivity. Due to the scarcity of polonium in nature and its rapid disintegration, even Marie Curie herself had expressed doubt at the beginning as to whether it could be an element since the spectroscopic evidence was inconclusive. It also took decades before it was possible to produce polonium in sufficient

[6] Nobel Archives, The Royal Swedish Academy of Sciences, Stockholm, Minutes of the Nobel proceedings of the Royal Academy of Sciences 1911, nominations by G. Darboux, dated 25 September 1910 and by S. Arrhenius, both registered on 28 January 1911.

amounts to investigate it — even X-ray spectroscopical evidence was not available until 1938. Moreover, there is nothing to suggest that Marie Curie's utmost goal — to determine the atomic weight of polonium in the traditional chemical analytical way — was ever accomplished before the new era of nuclear science and atomic numbers to identify chemical elements took over. Above all, the initial doubt and lack of proof indicate that the discovery process is far from being clear-cut and is rather messy and entangled.

Indeed, in the era of radioactivity, few of the then-existing "rules" to prove a scientific discovery applied. Instead of working with physical samples that could be analysed chemically to identify their components, the electrometer was used to detect the effect of emitted rays. This method could be used on "mixed" samples, where the elements in question were hidden somewhere inside; that is, a sort of chemistry of the "imponderable", as Marie Curie expressed in her Nobel lecture in 1911. However, chemical characteristics came in handy to separate the elemental fraction as much as possible from other elements in the mixture, ideally ending with determining what was considered the unique property of an element at the time — its atomic weight. The final, ideal proof, if at all reachable, was the isolation of the element in its elementary form, in this case, as a pure metal. How far did the chemistry of the imponderable then extend? Indeed, we must conclude that when the step of atomic weight determination was reached, although the rays themselves were imponderable and the methods and materials were no longer of that nature, they were as ponderable as could be, and Marie Curie and her team knew that even in the realm of imponderables, only the ponderable proof could convince chemists. 1898, therefore, constituted to be only one of the several steps in the discovery process of polonium and radium.

## Acknowledgements

The author thanks Brigitte Van Tiggelen and two external reviewers for their valuable feedback on the manuscript.

## References

Adloff, J.-P. (2011). A short history of polonium. *Chemistry International*, January–February, pp. 20–23.

Adloff, J.-P. and Kauffman, G. B. (2007). The troubled story of polonium: An early controversy in radiochemistry. *The Chemical Educator,* **12**, pp. 94–101.

Allchin, D. (2003). Science-myth conceptions. *Science Education*, **87**, pp. 329–351.

Badash, L. (1966). The discovery of thorium's radioactivity. *Journal of Chemical Education,* **43**, pp. 219–220.

Badash, L. (1979). *Radioactivity in America. Growth and Decay of a Science* (Johns Hopkins University Press, Baltimore, MD, USA and London, UK).

Bagnall, K. W. (1957). The chemistry of polonium. *Quarterly Reviews, Chemical Society,* pp. 30–48.

Bagnall, K. W. (1962). The chemistry of polonium. *Advances in Inorganic Chemistry and Radiochemistry*, **4**, pp. 197–229.

Bagnall, K. W. (1963). Polonium. *Endeavour,* **22**, pp. 61–64.

BNA. (2013). Marie Curie isolates pure radium metal — 7 September 1910. Published 6 September 2013 in The British Newspaper Archive: https://blog.britishnewspaperarchive.co.uk/2013/09/06/marie-curie-isolates-pure-radium-metal-7-september-1910/ (last accessed 25 January 2023).

BnF. Bibliothèque nationale de France, Département des Manuscrits. NAF18380. Pierre et Marie Curie. Papiers. I — OEUVRES ET TRAVAUX SCIENTIFIQUES. XV–XVIII Pierre et Marie Curie. Travaux sur la découverte de la radioactivité naturelle. XV–XVII Carnets de laboratoire, dits «Carnets de la découverte» du polonium et du radium. 1897–1900. Available from gallica.bnf.fra.

Bonolis, L. (n.d.). Prof. Dr. Irène Joliot-Curie research profile. *Lindau Nobel Laureate Meetings* (https://mediatheque.lindau-nobel.org/laureates/joliot-curie/research-profile, last accessed 26 January 2023).

Boudia, S. (1997). The Curie laboratory: Radioactivity and metrology, *History and Technology,* **13**, pp. 249–265.

Boudia, S. (2001). *Marie Curie et son laboratoire. Sciences et industrie de la radioactivité en France* (Éditions des archives contemporaines, Paris).

Brush, S. G. (1974). Should the history of science be rated X? *Science*, **183**, pp. 1164–1172.

Curie, I. (1925). Extraction et purification du dépot a évolution lente du radium. *Journal de chimie physique,* **22**, pp. 471–487.

Curie, M. (1898). Rays emitted by the compounds of uranium and thorium [translation of article in *Comptes rendus de l'Académie des Sciences,* **126**, pp. 1101–1103] in Romer. A. *Radiochemistry and the Discovery of Isotopes* (Dover Publications Inc., New York, 1970), pp. 65–68.

Curie, M. (1902–03). On the radioactive substance "polonium" [translation of article in *Physikalische Zeitschrift,* **4**, pp. 234–235] in Romer, A. *Radiochemistry and the Discovery of Isotopes* (Dover Publications Inc., New York, 1970), pp. 85–87.

Curie, M. (1911). Radium and the new concepts in chemistry. Nobel lecture (https://www.nobelprize.org/prizes/chemistry/1911/marie-curie/lecture/, last accessed 10 February 2021).

Curie, M. (1923/1963/2012). *Pierre Curie. With Autobiographical Notes by Marie Curie* (Dover Publications, Inc. Mineola, New York).

Curie, M. and Debierne, A. (1910). Sur le polonium. *Le Radium*, **7**, pp. 38–40.

Curie, P. and Curie, M. (1898). On a new radio-active substances contained in pitchblende [translation of article in *Comptes rendus de l'Académie des Sciences,* **127**, pp. 175–178] in Romer. A. *Radiochemistry and the Discovery of Isotopes* (Dover Publications Inc., New York, 1970), pp. 68–71.

Curie, P, Curie, Mme P. and Bémont, G. (1898). On a new, strongly radioactive substances contained in pitchblende [translation of article in *Comptes rendus de l'Académie des Sciences,* **127**, pp. 1215–1217] in Romer, A. *Radiochemistry and the Discovery of Isotopes* (Dover Publications Inc., New York, 1970), pp. 71–74.

Curie, P. and Curie, M. S. (1902). Sur les corps radioactifs. *Comptes rendus de l'Academie des Sciences*, **134**, pp. 85–87.

Demarçay, E. (1898). On the spectrum of a radio-active substance [translation of article in *Comptes rendus de l'Académie des Sciences,* **127**, p. 1218] in Romer. A. *Radiochemistry and the Discovery of Isotopes* (Dover Publications Inc., New York, 1970), pp. 74–75.

Des Jardins, J. (2011). Madame Curie's passion. *Smithsonian Magazine*, October 2011 (https://www.smithsonianmag.com/history/madame-curies-passion-74183598/, last accessed 26 January 2023).

Fernelius, W. C. (1982). Polonium. *Journal of Chemical Education*, **59**, pp. 741–742.

Fontani, M., Costa, M. and Orna, M. V. (2015). *The Lost Elements: The Periodic Table's Shadow Side* (Oxford University Press).

Giesel, F. (1899). Remarks on the behavior of radioactive barium and on polonium [translation of article in *Annalen der Physik*, **69**, pp. 91–94] in Romer, A. *Radiochemistry and the Discovery of Isotopes* (Dover Publications Inc., New York, 1970), pp. 77–79.

Hemmungs Wirtén, E. (2015a). *Making Marie Curie: Intellectual Property and Celebrity Culture in an Age of Information* (Chicago University Press).

Hemmungs Wirtén, E. (2015b). The pasteurization of Marie Curie: A (meta)biographical experiment, *Social Studies of Science*, **45**, pp. 597–610.

Hochwalt, C. A. and Haring, M. M. (1948). Manhattan District History. Book VIII, Los Alamos Project (Y) — Volume 3, Auxiliary Activities, Chapter 4, Dayton Project (https://www.osti.gov/includes/opennet/includes/MED_scans/Book%20VIII%20-%20%20Volume%203%20-%20Auxiliary%20Activities%20-%20Chapter%204,%20Da.pdf, last accessed 6 November 2022).

Hulubei, H., Cauchois, Y. and Cotelle, S. (1938). Déterminations spectroscopique du numéro atomique du polonium, *Comptes rendus de l'Académie des Sciences*, **207**, pp. 1204–1206.

Jacquemond, L.-P. (2019). Irène Joliot-Curie and the discovery of "artificial radioactivity". In *Women in their Element: Selected Women's Contributions to the Periodic System* (World Scientific, Singapore), pp. 361–373.

Malley, M. C. (1976). From hyperphosphorescence to nuclear decay: A history of the early years of radioactivity. 1896–1914, PhD thesis, University of California, Berkeley.

Malley, M. C. (2011). *Radioactivity: A History of a Mysterious Science* (Oxford University Press, Oxford).

Marckwald, W. (1902–03). Das radioaktive Wismuth (Polonium). *Physikalische Zeitschrift*, **4**, pp. 51–54.

Mendeléeff, D. (1904). *An Attempt Towards a Chemical Conception of the Ether*. (London, New York and Bombay: Longmans, Green, and Co.)

Moyer, H. V. (1956). Survey of early operations. In *Polonium*, ed. Moyer, H. V. (United States Atomic Energy Commission, Technical Information Service Extension, Oak Ridge, Tennessee), pp. 1–6.

New York Times (1910). Pure radium obtained. *The New York Times*, 6 September 1910.

Nobelprize. (1903). The Nobel Prize in Physics 1903 (https://www.nobelprize.org/prizes/physics/1903/summary/, last accessed 27 January 2023).

Nobelprize. (1911). The Nobel Prize in Chemistry 1911 (https://www.nobelprize.org/prizes/chemistry/1911/summary, last accessed 27 January 2023).

Pais, A. (1986). *Inward Bound: Of Matter and Forces in the Physical World* (Clarendon Press, Oxford).

Parsons, E. (n.d.). Marie Curie, 7 November 1867 — 4 July 1934. Winner of two Nobel prizes and the first woman to ever win a Nobel. *New Scientist* (https://www.newscientist.com/people/marie-curie/, last accessed 26 January 2023).

Perrin, F. (1981). Joliot-Curie, Irène. In *Dictionary of Scientific Biography*, ed. Gillispie, C. C. (Charles Scribner's Sons, New York), Vol. 7, pp. 157–159.

Pycior, H. A. (1996). Pierre Curie and "his eminent collaborator M$^{me}$ Curie": Complementary partners. In *Creative Couples in the Sciences*, eds. Pycior, H. A., Slack, N. G. and Abir-Am, P. G. (Rutger University Press, New Brunswick, NJ), pp. 39–56.

Quinn, S. (1995). *Marie Curie: A Life* (Simon & Schuster, New York).

Reid, R. (1974). *Marie Curie* (Collins, London).

Rentetzi, M. (2019). "She is in the next room": Elizabeth Róna and polonium. In *Women in their Element: Selected Women's Contributions to the Periodic System* (World Scientific, Singapore), pp. 333–340.

Romer, A. (1970). *Radiochemistry and the Discovery of Isotopes* (Dover Publications Inc., New York).

Roqué, X. (1997). Marie Curie and the radium industry: A preliminary sketch, *History and Technology*, **13**, pp. 267–291.

Roqué, X. (2019). Marie Skłodowska Curie — polonium and radium. In *Women in their Element: Selected Women's Contributions to the Periodic System* (World Scientific, Singapore), pp. 259–268.

Rutherford, E. (1910). Properties of polonium. *Nature*, **82**, pp. 491–492.

Sethy, N. K., Sutar, A. K., Rath, P., Jha, V. N., Ravi, P. M. and Tripathi, R. M. (2015). A review of radio chemical analysis and estimation of $^{210}$Po in soil matrices. *Journal of Radiation Research and Applied Sciences*, **8**, pp. 590–596.

Soddy, F. (1904). Radio-tellurium, *Nature*, **69**, pp. 461–462.
Sopka, K. R. and Sopka, E. M. (2010). The Bonebrake Theological Seminary: Top-secret Manhattan Project site. *Physics in Perspective*, **12**, pp. 338–349.
Van Tiggelen, B. (2001). The discovery of new elements and the boundary between physics and chemistry in the 1920s and 1930s: The case of elements 43 and 75. In *Chemical Sciences in the 20$^{th}$ Century: Bridging Boundaries*, ed. Reinhardt, C. (Wiley-VCH, Weinheim).
Van Tiggelen, B. (2019). Spying on shy elements: Yvette Cauchois and her spectrograph. In *Women in their Element: Selected Women's Contributions to the Periodic System* (World Scientific, Singapore), pp. 246–256.
Wichers, E. (1950). Report of the Committee on Atomic Weights of the American Chemical Society. *Journal of the American Chemical Society*, **72**, pp. 1431–1432.

## Selected Bibliography

Adloff, J.-P. and Kauffman, G. B. (2007). The troubled story of polonium: An early controversy in radiochemistry, *The Chemical Educator*, **12**, pp. 94–101.
Allchin, D. (2003). Science-myth conceptions, *Science Education*, **87**, pp. 329–351.
Curie, M. (1911). Radium and the new concepts in chemistry (Nobel lecture https://www.nobelprize.org/prizes/chemistry/1911/marie-curie/lecture/, last accessed 10 February 2021).
Hemmungs Wirtén, E. (2015). *Making Marie Curie: Intellectual Property and Celebrity Culture in an Age of Information* (Chicago University Press).
Malley, M. C. (2011). *Radioactivity: A History of a Mysterious Science* (Oxford University Press, Oxford).
Quinn, S. (1995). *Marie Curie: A Life* (Simon & Schuster, New York).
Romer, A. (1970). *Radiochemistry and the Discovery of Isotopes* (Dover Publications Inc., New York).

# Predicting and Anticipating Discoveries of Elements and Isotopes

https://doi.org/10.1142/9789819813858_0006

# GERMANIUM — DISCOVERED AS THE PREDICTED *EKA*–SILICON?

Gisela Boeck

*Institut für Chemie, Universität Rostock*
*D-18051 Rostock, Deutschland*
*gisela.boeck@uni-rostock.de*

More than 150 years ago, the German chemist Lothar Meyer (1830–1895) and the Russian chemist Dmitrii Ivanovich Mendeleev (1834–1907) established systems that ordered most of the known chemical elements with respect to increasing atomic weight and simultaneously by taking their chemical properties into account. It turned out that the properties of the elements show a periodic dependence on their atomic weights. For that reason, the system was called periodic, and it was presented in different shapes and forms along with tabular representations called periodic tables.

At that time, however, only about 60 elements were known, as opposed to the 118 we know of today. Meyer and Mendeleev suspected the existence of more elements and left vacancies in their tables for those yet to be discovered. Meyer was quite cautious in his predictions about further elements — he used dashes to indicate places for undiscovered elements and only implicitly predicted the atomic weight for one of those in his first elements' arrangement for the future element germanium [Meyer, 1864, p. 137; Boeck, 2019;

Rocke, 2019]. In 1872, he used question marks for the unknown elements and estimated atomic weights, which were astonishingly accurate [Meyer, 1872, p. 301].

It is well-known that Mendeleev made more precise statements about the properties of the elements he expected to exist. In his 1871 paper on the periodic regularity of chemical elements, Mendeleev described three as-yet-undiscovered elements in detail and also commented on a few more [Mendeleev, 1871, pp. 196–206].

Among the elements predicted to exist, three — the so-called *eka*-boron, *eka*-aluminium and *eka*-silicon[1] — were discovered in the 20 years following the publication of his periodic system. Mendeleev considered these to be an experimental confirmation of his periodic law, as he had made predictions that came to be successful. In fact, these confirmations are still often presented as one of the main factors for the acceptance of the periodic system, even though historical research has demonstrated this was not the case.[2] Indeed, historians and philosophers have dwelled at length on these episodes to discuss the predictive power of the periodic system, then and now, but usually fail at carefully examining the circumstances and discussions surrounding the discovery of any of those three *eka*-elements.

In this chapter, I will focus on *eka*-silicon, the last element of this group of "predicted" elements to be discovered as germanium in 1886. Through careful analysis of the discussion surrounding its placement in the periodic system, I will show that the incentive to find a new element was not prompted by an urge to "fill the gap" in the periodic system. In fact, divergent proposals were made for germanium's place in the periodic system before it was identified as *eka*-silicon. Whereas this might seem to undermine the power of

[1]The use of the word *eka* will be explained in the section "Mendeleev's Table of 1871 and the Gaps".

[2]Compare for example [Kaji *et al.*, 2015]. Bernadette Bensaude-Vincent [2019] ascribes Mendeleev's success to two factors: his determination to distinguish between "element" and "simple substance" in the process of constructing the periodic system, and symmetrically, by reconfiguring the notion of the element as a relational entity within system. Bensaude-Vincent [1986] also emphasises the importance of formulating a natural law with no exceptions.

Mendeleev's prediction, it sheds light on how the periodic system was used and progressively consolidated. Indeed, the story about the discovery and positioning of germanium in the periodic system provides an interesting case study of an element discovered in a not-yet-stable conceptual framework.

## Mendeleev's Table of 1871 and the Gaps

The historical tables of chemical elements differ from those we are using today. Mendeleev's table from 1871 was presented in a short form (eight-column table). Figure 1 shows such an arrangement. If one reads the atomic weights from left to right, one may notice their increasing values (with an exception in the case of I and Te). Mendeleev characterised the groups designated with Roman numerals on the basis of the oxygen and hydrogen compounds of the respective elements as well as the acidic and basic characters. From a contemporary point of view, the short-form arrangement might seem strange. Especially unusual for today's chemists is the placement of main group elements and transition metals together in one group. In addition, some older element symbols are used in the representation, including the didymium, which later turned out not to be an element at all. The atomic weights are almost exclusively given as integers. Mendeleev used the terms *even* and *odd* series in correspondence to the number of the row [Mendeleev, 1871, pp. 152–153]. The members of the *even* series are placed on the left side in the group, with those of the *odd* series on the right side. Regarding Cu, Ag and Au, Mendeleev could not make a final decision because in some properties, they fit in with group I, while other properties fit better in group VIII. Hence, in Figure 1, they are shown both as part of group I and group VIII.

The table of 1871 included 63 elements and more than 30 blank spaces with dashes for still-unknown elements. While it differs from the periodic table we use today, it helps us understand how and why Mendeleev named these unknown elements. He wrote: "We can see, by referring to [Figure 1] ..., in which the periodic relations are

**Tabelle II.**

| Reihen | Gruppe I. — $R^2O$ | Gruppe II. — RO | Gruppe III. — $R^2O^3$ | Gruppe IV. $RH^4$ $RO^2$ | Gruppe V. $RH^3$ $R^2O^5$ | Gruppe VI. $RH^2$ $RO^3$ | Gruppe VII. RH $R^2O^7$ | Gruppe VIII. — $RO^4$ |
|---|---|---|---|---|---|---|---|---|
| 1 | H=1 | | | | | | | |
| 2 | Li=7 | Be=9,4 | B=11 | C=12 | N=14 | O=16 | F=19 | |
| 3 | Na=23 | Mg=24 | Al=27,3 | Si=28 | P=31 | S=32 | Cl=35,5 | |
| 4 | K=39 | Ca=40 | —=44 | Ti=48 | V=51 | Cr=52 | Mn=55 | Fe=56, Co=59, Ni=59, Cu=63 |
| 5 | (Cu=63) | Zn=65 | —=68 | —=72 | As=75 | Se=78 | Br=80 | |
| 6 | Rb=85 | Sr=87 | ?Yt=88 | Zr=90 | Nb=94 | Mo=96 | —=100 | Ru=104, Rh=104, Pd=106, Ag=108 |
| 7 | (Ag=108) | Cd=112 | In=113 | Sn=118 | Sb=122 | Te=125 | J=127 | |
| 8 | Cs=133 | Ba=137 | ?Di=138 | ?Ce=140 | — | — | — | — — — — |
| 9 | (—) | — | — | — | — | — | — | |
| 10 | — | — | ?Er=178 | ?La=180 | Ta=182 | W=184 | — | Os=195, Ir=197, Pt=198, Au=199 |
| 11 | (Au=199) | Hg=200 | Tl=204 | Pb=207 | Bi=208 | — | — | |
| 12 | — | — | — | Th=231 | — | U=240 | — | — — — — |

Figure 1: Mendeleev's 1871 arrangement of elements [Mendeleev, 1871, p. 151]. In the public domain.

shown that many elements are missing, and we confidently predict their discovery" [Mendeleev, 1871, p. 196].[3]

Mendeleev wanted to avoid the introduction of new names into science, so he decided to use the name of the nearest lower analogue of the odd elements (right side in a group) or even elements (left side in the group) in the same group and place a Sanskrit numeral (*eka, dvi, tri,* etc.) in front of the name. The unknown elements of group I (left part) would be called *eka*-caesium (Ec) — a failed prediction — and *dvi*-caesium (Dc). Later, francium would fit in this place. The missing element in the fourth row is *eka*-boron (same column with boron on the left side of group III). In the fifth row, two elements are missing. Because of their analogy to aluminium and silicon (same column with aluminium and boron on the right side of groups III and IV), they were called *eka*-aluminium and *eka*-silicon. These last three yet unknown elements were comprehensively described with regard to their expected properties: atomic weight, volatility, possible

[3]For all Mendeleev's quotations published in German, the translations of Jensen [2005] are used. Quotations from letters or papers of other scientists have been translated by the author.

compounds, etc., by Mendeleev in his 1871 paper [Mendeleev, 1871, pp. 196–206].

Mendeleev also discussed the question as to which minerals would contain *eka*-silicon. For instance, he thought that minerals containing titanium but also arsenic and tin were possible sources because of their close proximity to *eka*-silicon. He was convinced that for the other elements, the properties could be described by proceeding analogously. He wrote:

> The examples that I have given are sufficient to show how, by means of the periodic law, we can predict the properties of unknown elements; consequently I will not pursue this subject for the other missing elements in the system. The discoveries which would offer the most interest would be those of the following elements: *eka*-caesium, Ec = 175; *dvi*-cesium, Dc = 220; *eka*-niobium, En = 146; *eka*-tantalum, Et = 235; and the analogs of manganese: *eka*-manganese, Em = 100, and *tri*-manganese, Tm = 190. The absence of a whole series (the ninth) and nearly all of a large period (from Ce = 140) can hardly be an accident and is probably due to the nature of elements in question.
>
> [Mendeleev, 1871, p. 205][4]

In the following, he discussed the question if the number of elements is finite or unlimited.

## Predictions of Elements and Their Role in the Discovery Process

Recently, Sergio Sereno [2020] discussed the role of Mendeleev's predictions, stating that despite abundant literature on the predictivity of the periodic system, there were still a few open questions. He mentioned two general points of view: One can express the opinion that predictions have a high confirmatory value or that there is no internal

[4] Compare with the empty spots in Fig. 1.

connection between predictions and discoveries. For the second case, Sereno used the statement of Scerri and Worrall about the discovery process of gallium. They wrote: "[Paul Émile Lecoq] De Boisbaudran did *not* discover gallium as a result of testing Mendeleev's prediction. Instead, he operated quite independently by empirical means in ignorance of Mendeleev's prediction..." [Scerri and Worrall, 2001, p. 417]. De Boisbaudran (1838–1912), assuming that Mendeleev would claim the priority for the element, argued that gallium has a density of 4.7 (g/cm$^3$), which differed from Mendeleev's prediction of 5.5 (g/cm$^3$), but in a second determination, De Boisbaudran gave the value of 5.94 (g/cm$^3$)[5] for the density, which better fitted the prediction [Brock, 1992, pp. 206–207]. Later, de Boisbaudran emphasised that the discovery of gallium involved empirical techniques quite separate from anything related to Mendeleev's work [Boisbaudran, 1875, p. 1105]. Scerri and Worrall concluded that the prediction of *eka*-aluminium and its discovery came as a coincidence; gallium could have been discovered independently from Mendeleev's predictions and perhaps even without the knowledge of the periodic system [Scerri and Worrall, 2001, p. 417–418].

As for scandium (*eka*-boron), Akeroyd [2003] showed that Lars Fredrik Nilson (1840–1899) also discovered scandium independent of Mendeleev's predictions. Nilson actually made an estimate of the atomic weight that was incorrect, probably because he had isolated only a small amount of scandium. Per Teodor Cleve (1840–1905) later extracted larger amounts and corrected the atomic weight to 45, and it was he who announced it as Mendeleev's *eka*-boron. Akeroyd claimed that Cleve would have proceeded in exactly the same manner, completely independent of whether he knew the predictions of Mendeleev, and attempted to prove this using Cleve's experimental procedure [Akeroyd, 2003, p. 345]. However, the fact that Cleve explicitly referred to Mendeleev speaks against the claim [Cleve, 1879, p. 55].

Sereno regretted the lack of clear pronouncements in the literature about the discovery of germanium [Sereno, 2020, pp. 478–480].

---

[5] Nowadays, the density of germanium is given with the value 5,323 g/cm$^3$.

He referred to Scerri [2007, pp. 138–140], who shortly described the discovery process of this element and the assumption of Clemens Winkler (1838–1904) that it should be *eka*-antimony. As I will demonstrate, Mendeleev himself proposed that germanium was *eka*-cadmium, and this prediction also turned out to be wrong. The German chemist Victor von Richter (1841–1891) and Meyer suggested *eka*-silicon instead. The different suggestions are discussed in more detail below. This chapter will thus investigate more closely these successive assumptions on the place of germanium inside the periodic system, starting with Mendeleev's efforts to find *eka*-silicon, one of the missing elements in the fifth row (group IV, right column).

## Mendeleev's Input to the Discovery of Germanium

The first comprehensive paper on the discovery of germanium was given by Bonifatii Mikhailovich Kedrov [1965] more than 50 years ago. In the 21st century, not only Scerri but also Haustein [2004; 2011] have published about Winkler and germanium, as has Weinert [2020]. Kedrov's work is especially comprehensive and illuminates Mendeleev's attempts to find *eka*-silicon.

In his 1869 review in *Zeitschrift für Chemie*, Mendeleev wrote:

> It allows one to foresee the discovery of many new elements, e.g., analogues of Si and Al, with atomic weights between 65 and 75.
>
> [Mendeleev, 1869, p. 406]

Based on the comparison of properties, Mendeleev expected that *eka*-silicon would be found in compounds of titanium and zirconium, and he specifically mentioned the mineral perovskite [Richter, 1870, p. 991].[6] He immediately ordered minerals that he thought would contain these metals, and after receiving them from the mineralogist, Pëtr Arkad'evich Kochubei (1825–1892), he started to

---

[6] Perovskite is the mineral $CaTiO_3$. It was found by Gustav Rose (1798–1873) in the Ural Mountains.

analyse them. Documents in the Mendeleev archive confirm this effort to find *eka*-silicon in 1871 [Kedrov, 1965, pp. 17–20]. Mendeleev also checked the literature for new research on minerals with titanium. He discussed whether ilmenium,[7] announced in 1846 by Hans Rudolph Hermann (1805–1879), could be his *eka*-silicon. However, a few years earlier, in 1867, Jean Charles Galissard de Marignac (1817–1894) had criticised Hermann's work and stated that ilmenium was a mixture of titanium, niobium and tantalum [Fontani *et al.*, 2015, pp. 47–48].

Whether or not Mendeleev — after the discovery of gallium — came back to the investigations of minerals to find *eka*-silicon is difficult to confirm. Kedrov took this view, but the following quote seems to indicate Mendeleev was no longer working on the topic himself:

> We may hope that the discovery of *eka*-silicon, Es = 72 ($EsO_2$), whose presumed properties are described in Liebig's Annalen, will soon be achieved. It should be sought for especially in association with arsenic and titanium.
>
> [Mendeleev, 1875, pp. 971–972]

Kedrov claimed several times in his paper that Winkler — who finally discovered germanium, the predicted *eka*-silicon — was aware of Mendeleev's search. A number of Mendeleev's papers with predictions of new elements had been published in German journals, but whether Winkler studied these intensively during his time as a practical chemist in a blue pigment factory is improbable.

## Clemens Winkler — A Short Biography

Clemens Winkler (Figure 2) was born in what was called the "silver town" of Freiberg[8] near Dresden on 26 December 1838. His father, Kurt Alexander Winkler (1794–1862), had studied at the Freiberg

[7] The name "ilmenium" is a reference to the Ilmensky Mountains near Minsk.
[8] Rich silver ore deposits were discovered in the Middle Ages.

Figure 2: Clemens Winkler, 1875. Courtesy of Angela Kugler-Kießling, Department of Old Collection at University Library "Georgius Agricola" Freiberg.

Mining Academy in the Kingdom of Saxony. He worked in blue colour works and in smelting firms.[9] He was privileged to travel at state expense to Sweden to work with the famous (Jöns) Jacob Berzelius (1779–1848). When he came back to Freiberg, he held executive positions in local factories.

Clemens was the third son of the family. He attended schools in Freiberg and Dresden; he later studied at the Royal Trade School in Chemnitz, which was well known for its excellent education, especially in chemistry. From 1856 to 1857, he completed a practical year in blue colour works and then enrolled at the Freiberg Mining Academy. After finishing his studies, he wanted to continue his scientific career but respected his father's wish and went to work in a blue colour plant. He applied for a position in Oberschlema, and he later worked in another blue pigment plant in Niederpfannenstiel (Figure 3).[10] Winkler had to do a lot of practical work, which, however, did not exhaust his creative talent. He dealt with scientific

[9] Blue pigments were produced from cobalt-containing ore.

[10] Oberschlema and Niederpfannenstiel are small towns in the Ore Mountains in Saxony.

Figure 3: View of a blue pigment factory in Niederpfannenstiel in the middle of the 19th century. Courtesy of Mike Haustein.

issues that were mostly related to the concerns of the plant. For example, he conducted experiments to separate copper and cobalt from nickel [Haustein, 2004, pp. 22–31].

In this industrial period of Winkler's life, he published about 40 scientific papers. He was especially interested in the removal of sulfur dioxide from industrial emissions because he had seen the impact of sulfuric gases on the environment. He submitted two papers as a dissertation, one about silicon alloys and silicon arsenic metals and the other about the analytic determination of water in organic liquids, and he received his doctorate *in absentia* from the University of Leipzig in 1864 [Beyer, 2005, pp. 28–34].

In 1873, the Freiberg Mining Academy was looking for a professor of chemistry because Theodor Scheerer (1813–1875) had retired. Hermann Kolbe (1818–1884) from the university in Leipzig recommended the young metallurgical chemist Clemens Winkler. At the age of 34, Winkler was appointed professor of inorganic chemistry. Winkler remained in Freiberg for the rest of his scientific career, rejecting calls to other universities and offers for other professorships.

From 1896 to 1899, he was director of the Mining Academy. He died in Dresden on 8 October 1904. Winkler worked in varied fields, such as technical-chemical analysis of gases, analysis of a meteorite, and the improvement of sulfuric acid production. But I will concentrate here on his work in connection with germanium.

## Argyrodite — Germanium — *Eka*-Silicon

In September 1885, a previously unknown mineral was discovered in the Himmelsfürst mine outside Freiberg: grey, silver-rich with a coating of iron and pyrites on the outside (Figure 4). A sample of this

Figure 4: A container of argyrodite with the inscription: "Argyrodite from Clemens Winkler's collection, owner O. Brunck, a sample yielded 3.8 g", a few pieces of argyrodite, and the original sealed glass tube with $GeS_2$, initially obtained by Winkler on 6 February 1886. Courtesy of Technical University Minors' Academy Freiberg/Michael Schwan.[11]

[11]The Clemens-Winkler-Collection is part of the Collection of Inorganic Substances in the Institute of Inorganic Chemistry at Technical University Minors' Academy Freiberg, and the laboratory of Winkler is a historical landmark today. The collection items are currently being digitised and will be made available online. See https://winklersammlung.webvirt.hrz.tu-freiberg.de/

was given to the mineralogist Albin Weisbach (1833–1901). He recognised that this was a new mineral with a high silver content. For this reason, it was named "argyrodite" (meaning "rich in silver" in Greek). Theodor Richter (1824–1898), a specialist in blowpipe analysis and the co-discoverer of indium, determined that the silver content of the unknown mineral was 73.5 per cent; he also found amounts of sulfur and mercury in the mineral. Mercury was usually not present in the Freiberg ores. This may be the reason that Weisbach asked his cousin Winkler to verify the analysis of the mineral.

Winkler's first results indicated that the mineral indeed contained 73–75 per cent silver, 17–18 per cent sulfur, and small traces of mercury, iron and arsenic [Winkler, 1886a, 210]. Yet some of the minerals could not be identified. Although he had repeated the analyses several times, there was a glaring difference to 100 per cent. Winkler thought that this was not acceptable as an analytical result. He had already speculated that it might be a new element, and on 6 February 1886, his efforts led him to the sulfide of a new element, which he named "germanium" (Figure 4). The same day, he wrote a short note to *Berichte der Deutschen Chemischen Gesellschaft* (Reports of the German Chemical Society):

> The determination of the atomic weight of germanium should be carried out immediately, even if at first only with approximate accuracy; it will show whether, as may be assumed, the new element is called upon to fill the gap in the periodic system between antimony and bismuth.
>
> [Winkler, 1886a, p. 211]

The quotation shows that Winkler assumed that germanium could occupy the gap between antimony and bismuth, but he did not provide a value for the atomic weight and did not use the word *eka*-antimony that would correspond to Mendeleev's nomenclature [Winkler, 1886a]. In his first paper, Winkler only reported about the chemical properties of germanium. He had noted similarities to arsenic and antimony, so he believed that the new element belonged to the nitrogen group. Winkler had clearly seen Mendeleev's system (Figure 1) — a comparison with this figure shows that there was a gap in group V (right column) with a dash, which should symbolise an unknown element. However, to what extent he was familiar with Mendeleev's predictions remains uncertain. To my knowledge, Mendeleev did not refer to *eka*-antimony in any publication.

In any event, the search for an element, which had been predicted, was in no way the starting point for the investigation of argyrodite since it was not based on a suspicion that this new mineral would contain a new element. Only during the investigation process did Winkler understand that the parts that were unaccounted for had to be a new element.

In a letter dated 25 February 1886, von Richter wrote to Winkler that germanium could not be *eka*-antimony but that it must be *eka*-silicon instead. The expected properties of *eka*-antimony would, in fact, differ greatly from those exhibited by germanium. He furthermore added:

> *Eka*-silicon is the element we have been waiting for with the greatest excitement, and in any case the closer examination of the germanium will yield the most decisive *experimentum crucis* for the periodic system. It would be a nice coincidence if germanium (Ge = 72.5) were to stand next to gallium (Ga = 69.8).[12]

Interestingly, von Richter provided an atomic weight here. As discussed above, this was not provided in Winkler's first paper. I assume that there were more letters between the two scientists than are kept in the archives, but they have not survived.

In Winkler's collection, there is also a postcard that Winkler received from von Richter on 12 or 28 February 1886.[13] The postmark is difficult to decode. Since Richter congratulated Winkler on

---

[12] Department of Old Collection at University Library "Georgius Agricola" Freiberg, Clemens Winkler's collection, a letter of Richter to Winkler, 25 February 1886.

[13] Department of Old Collection at University Library "Georgius Agricola" Freiberg, Clemens Winkler's collection, a postcard of Richter to Winkler.

Figure 5: Both sides of von Richter's postcard to Winkler. Courtesy of Angela Kugler-Kießling, Department of Old Collection at University Library "Georgius Agricola" Freiberg.[14]

the discovery of germanium in the letter of 25 February, it can be assumed that the postcard was written afterwards (Figure 5). Possibly, it was a response to an enquiry from Winkler about Mendeleev's predictions because Winkler was not very well acquainted with the periodic law. However, Winkler mentioned the possibility that the new element was Mendeleev's *eka*-antimony in his first paper, which shows that he must have known about Mendeleev's periodic system (Figure 1).

Von Richter recommended Meyer's book *Modern Theories of Chemistry and their Significance for Chemical Statics* [Meyer, 1864] as a source of information about the periodic system, along

[14] (Asterisks mark uncertain translations since the handwriting is difficult to read in certain places.) Dear Mr. Colleague! You will find information about the periodic system in L. Meyer's "Modern Theories", however, simpler and clearer, as I believe, in Richter's textbook of inorganic chemistry (from the first edition on). The only summarising treatise of Mendelejeff is to be found in *Liebig Ann. Sup. VII*, p. 133 in said place (p. 201) about the properties of *Eka*silicon. These predictions *had to be* modified several times after the discovery of gallium. Tomorrow, I will send a short communication to the *Berichte*, in which I point out the importance of the discovery of germanium and the law of periodicity. I will also send you a copy of it at the same time. Yours sincerely, V. V. Richter.

with his own textbook [Richter, 1875] and Mendeleev's work in *Liebigs Annalen* [Mendeleev, 1871]. In fact, he explicitly mentioned page 201, where Winkler might find the predicted properties of *eka*-silicon.

Also, Lothar Meyer read Winkler's paper in the *Berichte* and showed an interest in germanium. He started to write a short note and subsequently sent a letter to Winkler (likely dated 27 February 1886):

> This morning, I had presented and justified to my older students the opinion that your new element is *eka*-silicon and now I wanted to point out in *Berichte der chemischen Gesellschaft* that Mendelejeff had been as unlucky with the prediction of the properties as with gallium. As infallible as M. considers himself to be in these matters, he has made the rather gross violation of stamping the *eka*-silicon as its homologues of titanium and thus declaring it to be refractory. He could have gathered from my atomic volume curve already at that time that it was analogous to tin and consequently easily fusible and almost as volatile as its neighbour arsenic.[15]

Meyer refers here to his "paper tool" [Klein, 2003], the atomic volume curve, which he introduced in 1870, and which showed the periodicity of the elements properties in an impressive way [Meyer, 1870]. However, Meyer was wrong about the ease of fusing.[16]

Mendeleev himself sent a letter to Winkler, dated 26 February 1886. He congratulated Winkler on the discovery of germanium and underlined the high interest in answering the question as to which place this new element should occupy in the periodic table. Mendeleev offered the following place for germanium:

> If it is correct that the place of germanium is found between Sb and Bi, its atomic weight must be equal to 160–165 ... However, it

[15] Department of Old Collection at University Library "Georgius Agricola" Freiberg, Clemens Winkler's collection, a letter from Meyer to Winkler, 27 February 1886.

[16] For further discussion, see [Boeck and Rocke, 2022, p. 29].

> seems to me that the solubility of germanium chloride in water and the characteristic white colour of germanium sulfide point to the possibility of a different position of germanium and make it desirable to try to define it as *eka*-cadmium. For this element, whose position is between Cd and Hg, the atomic weight must be about 155 ... The great volatility of germanium itself and the certain volatility of its chloride do not allow us to consider it to be *eka*-silicon, although other properties are quite close.[17]

Over the course of the next months, Winkler started to test the properties of germanium and to determine its atomic weight. This analytical work was quite difficult because the germanium compounds were easily soluble and highly volatile. The other problem was that Winkler did not have enough raw material, meaning he lacked argyrodite. Most of the ore had already been processed in the smelter. But eventually, he got about 500 kg of an ore containing 50–60 per cent argyrodite. After purification, he was left with 156 g germanium for analysis in his laboratory (and to use as gifts for colleagues). Winkler determined the atomic weight to be 72.32. He could also show that the valency of germanium was 4 because germanium formed a chloride in the form $GeCl_4$, as was shown by the determination of the vapour density. Consequently, it belongs to the carbon group in the periodic system [Brunck, 1906, pp. 4532–4533].

During this period of intensive analytical work, Winkler exchanged several letters with Mendeleev. He informed him that the experiments confirmed the fact that germanium definitely was *eka*-silicon [Kedrov, 1965, pp. 26–27]. Winkler summarised his results in a comprehensive publication in the summer of 1886. He stated that in his first note on germanium, he had expressed the assumption that the new element would be *eka*-antimony because there were several analogies to antimony and arsenic. But other properties and the atomic weight were still unknown, and he warned against

[17] Department of Old Collection at University Library "Georgius Agricola" Freiberg, Nachlass C. Winkler, a letter from Mendeleev to Winkler, dated 26 February 1886.

careless analogies. This case demonstrates how deceptive it can be to rely on analogies, which the author insisted upon [Winkler, 1886b, p. 182]. Winkler also stressed that Meyer immediately recognised germanium as *eka*-silicon because his atomic volume curve showed the low melting and vaporisation points, but then Winkler expressed the importance of verifying Mendeleev's prediction; he wrote:

> There can hardly be stronger proof of the doctrine of the periodicity of the elements than the materialization of the so far hypothetical *eka*-silicon, and it forms in truth more than the mere confirmation of a bold theory, it means an eminent extension of the chemical horizon, a mighty step into the realm of knowledge.
>
> [Winkler, 1886b, p. 183]

After reading Winkler's paper of July, Meyer stated that he was deeply impressed that Winkler never lost the idea of a new element regardless of the complicated experimental conditions. He wrote that another proper chemist would have thrown in the towel on the matter and had explained the missing per cent in the analysis with a breezy hypothesis. He also remembered the fact that the German chemist Justus von Liebig (1803–1873) had failed to discover bromine.[18] Meyer was acquainted with the history that Liebig was requested to analyse the salts and brines of a little town near Kreuznach. However, his analyses were "sloppy". He misinterpreted bromine as an unrecognised chloride of iodine [Brock, 1997, p. 117].

Many scholars celebrated the confirmation of Mendeleev's prediction, first of all Mendeleev himself. For him, Winkler was one of the verifiers of the periodic system. That is why Mendeleev hung Winkler's picture above his desk next to the photographs of the discoverers of gallium and scandium, as these three chemists confirmed the predictions about the existence and, for the most part, the

[18] Department of Old Collection at University Library "Georgius Agricola" Freiberg, Nachlass C. Winkler, a letter from Meyer to Winkler, dated 30 August 1886.

Figure 6: Portraits of the four foreign scientists — fortifiers of the Periodic Law (Lars Fredrik Nilson is on the left while Clemens Winkler is on the right. Paul Émile Lecoq de Boisbaudran is at the top while Bohuslav Brauner is at the bottom) © St. Petersburg State University, Charin Ivan, D. I. Mendeleev Museum-Archive of St. Petersburg State University, registration number: MM KP-260.

properties of these elements and thus contributed to Mendeleev's fame (Figure 6).[19]

***

Mendeleev predicted several elements, three of which were discovered during his lifetime. Not only did he predict their existence, but he also described their expected properties. Immediately after the description of *eka*-boron, *eka*-aluminium and *eka*-silicon, Mendeleev looked for *eka*-silicon in minerals, but he gave up this work shortly after. The search for the three elements that were discovered was, however, not done with the goal of confirming the predictions or

[19]The photo of Bohuslav Brauner (1855–1935) is placed in this collection because he was a great supporter of Mendeleev's periodic system; compare [Štrbáňová, 2015].

using the periodic law as a guide. Germanium was initially identified by Winkler himself as *eka*-antimony, one of Mendeleev's failed predictions. Mendeleev later on suggested its identification with *eka*-cadmium, which was another of his predictive failures. Meyer and von Richter simultaneously identified the new element correctly as *eka*-silicon. In this case, the existence of predictions of unknown elements did not help the discovery and even hindered the correct identification of the new element. At the beginning of the 20th century, the atomic theory explained the theoretical basis for the periodic system. From this moment on, it became a tool and a compass for a targeted search for new elements, including the heavy elements produced in accelerators.[20] The case of germanium also shows that the discovery process can be slowed by other more tangible factors, for instance, the availability of raw material and the chemists' analytical capability, especially their striving for precision and their perseverance. From the point of view of the chemical community, the story of germanium illustrates the intense and fruitful communications taking place among chemists at the end of the 19th century and gives some insight into the relationships between the two most important discoverers of the periodic system and their understanding of how the system would work as a predictive tool.

The negotiation of germanium's placement in the periodic system does, however, demonstrate two important points: there was an impulse to situate the new-found element within that framework, and this prompted a negotiation between the periodic system and the experimental finding as to germanium's identity. This led to a better definition of germanium both as an element bearer of specific qualities and as *eka*-silicon; that is, an element in relation to neighbouring elements, including the family formed by the carbon group. Furthermore, whereas the discovery of germanium was not prompted by Mendeleev's detailed predictions for three *eka*-elements, it did become a confirmation of the periodic system's ability to organise all elements in a natural and rational way. Ironically, it was Meyer's atomic volume curve that provided the best extrapolation for the

[20] See also Brigitte Van Tiggelen's and Ann Robinson's contributions to this volume.

properties, even though Meyer had always been publicly very reserved about making any detailed predictions. Germanium only became *eka*-silicon in the course of the process, and the periodic system was, in any event, not the starting point for its discovery.

## Acknowledgements

The author wishes to thank Mike Haustein, Edwin Kroke, Institute of Inorganic Chemistry, Technical University Mining Academy Freiberg, Angela Kugler-Kießling, Department of Old Collection at University Library "Georgius Agricola" Freiberg, and Renée Lüskow in Rostock for their support in this work. My thanks also go to the reviewers for their valuable advice. I am very grateful to the editors of the volume for the invitation to publish this paper and for their everlasting help.

## References

Akeroyd, F. M. (2003). Prediction and the periodic table: A response to Scerri and Worrall, *Journal for General Philosophy of Science*, **34**(2), pp. 337–355.

Bensaude-Vincent, B. (1986). Mendeleev's periodic system of chemical elements, *The British Journal for the History of Science,* **19**(1), pp. 3–17.

Bensaude-Vincent, B. (2019). Reconceptualizing chemical elements through the construction of the periodic system, *Centaurus*, **61**(4), pp. 299–310.

Beyer, L. (2005). *Vom Doktoranden zum bedeutenden Chemiker* (Passage-Verlag, Leipzig).

Boeck, G. (2019). Julius Lothar (von) Meyer (1830–1895) and the periodic system, *Substantia,* **3**(2), suppl. 4, pp. 13–25.

Boeck, G. and Rocke, A. J. (2022). *Lothar Meyer: Modern Theories and Pathways to Periodicity* (Birkhäuser, Basel).

Boisbaudran, P. É. Lecoq de (1875). Sur quelques propriétés du gallium. *Comptes rendus*, **81**, pp. 1100–1105.

Brock, W. H. (1992). *The Fontana [Norton] History of Chemistry* (London: Fontana, 1992; New York: Norton, 1992).

Brock, W. H. (1997). *Justus von Liebig — The Chemical Gatekeeper* (Cambridge University Press, Cambridge, New York, Melbourne).

Brunck, O. (1906). Clemens Winkler, *Berichte der Deutschen Chemischen Gesellschaft*, **39**, pp. 4491–4548.

Cleve, T. (1879). Sur le scandium, *Comptes Rendus*, **89**, pp. 419–422.

Fontani, M., Costa, M. and Orna M. V. (2015). *The Lost Elements — The Periodic Table's Shadow Side* (Oxford University Press).

Haustein, M. (2004). *Clemens Winkler: Chemie war sein Leben* (Verlag Harry Deutsch, Frankfurt am Main).

Haustein, M. (2011). Germanium, *Chemie in unserer Zeit*, **45**(6), pp. 398–405.

Jensen, W. B. (2005). *Mendeleev on the Periodic Law* (Dover, Mineola, NY).

Kaji, M, Kragh, H. and Palló, G. (2015). *Early Responses to the Periodic System* (Oxford University Press, New York).

Kedrov, B. M. (1965). Die Vorhersage des Ekasiliziums und die Entdeckung des Germaniums, *NTM Schriftenreihe für Geschichte der Naturwissenschaften, Technik und Medizin*, 2(8), pp. 11–37.

Klein, U. (2003). *Experiments, Models, Paper Tools: Cultures of Organic Chemistry in the Nineteenth Century* (Stanford University Press, Stanford).

Mendeleev, D. I. (1869). Über die Beziehung der Eigenschaften zu den Atomgewichten der Elemente, *Zeitschrift für Chemie*, pp. 405–406.

Mendeleev, D. I. (1871). Die periodische Gesetzmäßigkeit der chemischen Elemente, *Annalen der Chemie und Pharmacie*, **8**, pp. 133–229.

Mendeleev, D. I. (1875). Remarques à propos de la découverte du gallium, *Comptes rendus hebdomadaires des séances de l'Académie des Sciences*, **81**, pp. 969–972.

Meyer, L. (1864). *Die modernen Theorien der Chemie und ihre Bedeutung für die chemische Statik* (Verlag von Maruschke & Berendt, Breslau).

Meyer, L. (1870). Die Natur der chemischen Elemente als Function ihrer Atomgewichte, *Annalen der Chemie und Pharmacie*, 7(1870), pp. 354–364.

Meyer, L. (1872). *Die modernen Theorien der Chemie,* 2$^{nd}$ ed. (Verlag von Maruschke & Berendt, Breslau).

Richter, V. von (1870). Correspondenz aus Petersburg 6/18. Decbr., *Berichte der Deutschen Chemischen Gesellschaft*, **3**, pp. 988–992.

Richter, V. von (1875). *Lehrbuch der anorganischen Chemie* (Verlag von Friedrich Cohen, Bonn).

Rocke, A. J. (2019). Lothar Meyer's pathway to periodicity, *Ambix*, **66**(4), pp. 265–302.
Scerri, E. (2007). *The Periodic Table, Its Story and Its Significance* (Oxford University Press, New York).
Scerri, E. R. and Worrall, J. (2001). Prediction and the periodic table, *Studies in History and Philosophy of Science*, **32**(3), pp. 407–452.
Sereno, S. G. M. (2020). Prediction, accommodation and the periodic table: A reappraisal, *Foundations of Chemistry*, **22**(3), pp. 477–488.
Štrbáňová, S. (2015). Nationalism and the process of the reception and appropriation of the periodic system in Europe and the Czech Lands. In Kaji, M, Kragh, H., and Palló, G. (2015), pp. 121–149.
Weinert, Ch.S. (2020). Die Chemie ist schwierig: Winkler and the Discovery of Germanium, *Bulletin for the History of Chemistry*, **45**, pp. 8–15.
Winkler, C. (1886a). Germanium, Ge, ein neues, nichtmetallisches Element, *Berichte der Deutschen Chemischen Gesellschaft*, **19**, pp. 210–211.
Winkler, C. (1886b). Mitteilungen über das Germanium, *Journal für praktische Chemie*, **34**, pp. 177–229.

## Selected Bibliography

Boeck, G. and Rocke, A. J. (2022), *Lothar Meyer: Modern Theories and Pathways to Periodicity* (Birkhäuser, Basel).
Fontani, M., Costa, M. and Orna M. V. (2015), *The Lost Elements — The Periodic Table's Shadow Side* (Oxford University Press).
Gordin, M. D. (2019). *A Well-Ordered Thing: Dmitrii Mendeleev and the Shadow of the Periodic Table*, rev. ed. (Princeton University Press, Princeton).
Haustein, M. (2004). *Clemens Winkler: Chemie war sein Leben* (Verlag Harry Deutsch, Frankfurt am Main).
Rocke, A. J. (2019). Lothar Meyer's pathway to periodicity, *Ambix*, **66**(4), pp. 265–302.
Scerri, E. (2007). *The Periodic Table, Its Story and Its Significance* (Oxford University Press, New York).

https://doi.org/10.1142/9789819813858_0007

# IT'S NOT ELEMENTARY: THE SUCCESSIVE LIVES OF *EKA*-MANGANESE AKA ELEMENT 43

Brigitte Van Tiggelen

*Mémosciences asbl, Voie du Vieux Quartier 18,*
*1348 Louvain-la-Neuve, Belgium*
*Science History Institute, 315 Chestnut Street,*
*Philadelphia, PA 19106, USA*
*bvantiggelen@sciencehistory.org*

Short narratives in chemistry textbooks, encyclopaedias and popular science books devoted to the discovery of a chemical element often display a standardised format [Hackney, 1951; Murin *et al.*, 1961; Ball, 2021]: Before getting to the description and properties of the element and its compounds, the main pieces of information provided are often when, where, how and by whom an element was first observed, isolated, and claimed to be discovered. Depending on the space allotted to the historical development, ensuing controversies are also included. These short accounts may also mention "forerunners" that appeared before the actual consensus emerged among the community. In that case, "forerunners" designate both scientists and substances that can qualify for playing a part in the progressive accumulation of knowledge that led to the new element being established as such. Indeed, the term covers not only those who claimed to have made the discovery of a new element and whose claim was later rejected, but it also denotes those substances that were alleged to be new chemical elements but were never fully confirmed nor infirmed, including some dismissed elements for instance that were shown to be a mixture of known elements.[1]

These stories express two important features. First, they see past attempts as part of the "biography" of the elements, as science writer Philip Ball presents this in his book on chemical elements [Ball, 2021]. Second, they often aim at fairness by listing some of the previous attempts and exhuming figures that do not appear in a triumphant story of element discovery. Written from a presentist point of view, however, many of these stories fail to do full justice to the original actor's perspective by ignoring the horizon of possibilities and the limits of prediction available to those who made these claims. Most of the retrospective identifications of discovery claims do not provide the context, both practical and theoretical, in which these claims were made but only seek to identify the material through the physical and chemical description of the properties and processes. What is fundamentally missing is what I will call the "horizon of expectations": What were the actors thinking they were doing, and what did they believe were the potential outcomes of their manipulations?

In the case of elements, what was these "discoverers'" concept of elements, and what were the forecasted properties and mode of existence they were anticipating? From a historical point of view, failing to address such questions is falling into anachronism, when one sketches an event of the past without considering the context of the time to make sure it is not interpreted with a presentist or 'Whiggish' perspective. The aim of this contribution, however, is not to point at such historical "mistakes", but, on the contrary, to understand what drives this way of thinking that is so frequent in discovery narratives when it comes to elements. In fact, this

[1] See also Charlotte Abney Solomon's contribution to this volume.

particular feature of historical narratives shared among chemists also reveals some interesting facets of the discovery process as viewed by the narrators of these stories and sheds light on the context-dependent narration. What it will also reveal is that through that process, writing the history of unveiling one single element always ends up being connected, loosely or not, with other elements.

This paper will focus on element 43, now known as technetium, Tc, and often declared as being predicted by the Russian chemist Dmitrii Ivanovich Mendeleev (1834–1907) as an element yet to be discovered. Indeed, in his 1871 table, Mendeleev indicated several places where, according to his periodic law, elements ought to exist and hence were waiting to be discovered (Figure 1). These elements-to-be received provisional names, depending on their positioning in a traditional representation of his periodic system with the groups arranged vertically: the prefix *eka-* designated the element standing or was assumed to stand beneath or under a specified element in that group. The next in the row would be *dvi-*, then *tri-*, and so on, according to the first numerals in Sanskrit. Among others, Mendeleev

**Tabelle II.**

| Reihen | Gruppe I. — $R^2O$ | Gruppe II. — $RO$ | Gruppe III. — $R^2O^3$ | Gruppe IV. $RH^4$ $RO^2$ | Gruppe V. $RH^3$ $R^2O^5$ | Gruppe VI. $RH^2$ $RO^3$ | Gruppe VII. $RH$ $R^2O^7$ | Gruppe VIII. — $RO^4$ |
|---|---|---|---|---|---|---|---|---|
| 1 | H=1 | | | | | | | |
| 2 | Li=7 | Be=9,4 | B=11 | C=12 | N=14 | O=16 | F=19 | |
| 3 | Na=23 | Mg=24 | Al=27,3 | Si=28 | P=31 | S=32 | Cl=35,5 | |
| 4 | K=39 | Ca=40 | —=44 | Ti=48 | V=51 | Cr=52 | Mn=55 | Fe=56, Co=59, Ni=59, Cu=63 |
| 5 | (Cu=63) | Zn=65 | —=68 | —=72 | As=75 | Se=78 | Br=80 | |
| 6 | Rb=85 | Sr=87 | ?Yt=88 | Zr=90 | Nb=94 | Mo=96 | —=100 | Rn=104, Rh=104, Pd=106, Ag=108 |
| 7 | (Ag=108) | Cd=112 | In=113 | Sn=118 | Sb=122 | Te=125 | J=127 | |
| 8 | Cs=133 | Ba=137 | ?Di=138 | ?Ce=140 | — | — | — | — — — — |
| 9 | (—) | — | — | — | — | — | — | |
| 10 | — | — | ?Er=178 | ?La=180 | Ta=182 | W=184 | — | Os=195, Ir=197, Pt=198, Au=199 |
| 11 | (Au=199) | Hg=200 | Tl=204 | Pb=207 | Bi=208 | — | — | |
| 12 | — | — | — | Th=231 | — | U=240 | — | — — — — |

Figure 1. Mendeleev's 1871 arrangement of elements [Mendeleev, 1871, p. 151]. The predicted atomic weight of 100 is shown for an expected homologue of Mn, in group VII, situated on row 5, between molybdenum and ruthenium. In the public domain.

left two gaps for unknown elements with similar properties to manganese and higher atomic weights, which were called *eka*- and *dvi*-manganese.

Mendeleev even indicated an estimated atomic weight of 100 for the forecasted *eka*-manganese, but contrary to *eka*-boron, *eka*-aluminum, and *eka*-silicon,[2] he did not provide any further detailed prediction of the physical properties and chemical behaviour besides the interpolated atomic weight. For *dvi*-manganese, there was not even any mention of atomic weight estimation. Mendeleev stopped providing atomic weights for *eka*-manganese in later publications. And, in fact, chemists struggled for some time to isolate what was provisionally called *eka*-manganese. In hindsight, we know the chemistry of the elements later known as transition metals, and especially those close to the platinum family is quite delicate and requires very fine and meticulous craftmanship that only developed in the course of time. Confusions with neighboring elements bearing analogous properties were therefore inevitable and indeed numerous. This is particularly the case with *dvi*-manganese, which pops up several times in the discovery story of *eka*-manganese.

Technetium, textbooks teach us, is the lighter, unstable element of the periodic table, with all of its isotopes undergoing radioactive decay. Its longest-lived isotope is much younger than the earth, but technetium can, however, be found on earth in minute quantity as a fission product of uranium. It was also observed in stellar spectral lines, but these observations were only made after men had to produce it with nuclear physics instruments such as cyclotrons. This is, in fact, how it was first acknowledged after being produced artificially when, in 1937, the Italian physicists Carlo Perrier (1886–1948) and Emilio Segrè (1905–1989) chemically separated a new element from a molybdenum foil that had been bombarded by neutrons or deuterons (or a hydrogen nucleus) in the Lawrence Berkeley National Laboratory in California. In 1947, the name "technetium" from the Greek word τεχνητός, meaning "artificial", was adopted for element

[2] See Gisela Boeck's contribution to this volume.

43 since it was the first element to be artificially produced [Perrier and Segrè, 1947].

Many claims were made for element 43 before it was accepted as technetium, and many names were proposed within the frame of the periodic system but not met with acceptance by the chemical community. When reviewing the history of element 43, several accounts mention elements that were claimed to be discovered and could be considered as forerunners of technetium, going back to even the pre-periodic system era, that is, before 1869. This chapter will analyse selected episodes in the context of the knowledge of the time, and scrutinise how these so-called discoveries were retrospectively assessed. This requires analysing both how a substance claimed to be newly observed in the past becomes identified as element 43 and how the finding has been deemed a discovery. Special attention will be given to the revisiting of historical narratives, endeavouring to demonstrate that predecessors to the acknowledged discoverer had indeed "observed" a new element even though they were unable to reproduce the piece of evidence needed at a level that brought them the recognition by the chemical community in their own time.

Instead of taking the chronological path, I have chosen to use a retrospective approach to unflesh this process of construction of an element's "biography". This will allow for a better measure of the increasing distance between the moment of the claim and the moment of reinterpretation of that same claim much later after the facts. It will clarify how the "horizon of expectations" was not only evacuated but also rebuilt by new interpretations in new contexts.

## Act I — Rehabilitating Masurium

In 1988, a 10-page contribution with the title "The ignored discovery of the element Z = 43" appeared in the journal *Nuclear Physics*, signed by the Belgian nuclear chemist Pieter Van Assche (1933–2022). In the first sentence of the abstract, the author based in Leuven, Belgium, claimed that Perrier and Segrè were wrongly credited with the first observation of the element with atomic number 43 (when they produced it artificially). In the second sentence, he stated that he was providing convincing arguments that the Berlin-based research team Walter Noddack (1893–1960), Ida Tacke (1896–1978), and Otto Berg (1873–1939), who had announced to have discovered the element along with element 75 back in 1925 should, in fact, be credited with its discovery. In their 1925 article, the team had declared to discover both elements 43 and 75, the heavier homologues of manganese, and suggested the name "masurium" and "rhenium", respectively, for these newcomers [Noddack and Tacke, 1925; Berg and Tacke, 1925]. While they were able to isolate a substantial sample of rhenium within five years after their announcement, the team — now restricted to Noddack and his wife[3] — was never able to provide the same evidence for the alleged discovery of masurium [Van Tiggelen, 2001].

Periodisches System der Elemente.

| 1 | 2 | 3 | 4 | 5 | 6 | 7 | 8 | 9 | 10 | 11 | 12 | 13 | 14 | 15 | 16 | 17 | 18 |
|---|---|---|---|---|---|---|---|---|---|---|---|---|---|---|---|---|---|
| | | | | | | | | | | | | | | | | 1 H | 2 He |
| 3 Li | 4 Be | | | | | | | | | | | 5 B | 6 C | 7 N | 8 O | 9 F | 10 Ne |
| 11 Na | 12 Mg | | | | | | | | | | | 13 Al | 14 Si | 15 P | 16 S | 17 Cl | 18 A |
| 19 K | 20 Ca | 21 Sc | 22 Ti | 23 V | 24 Cr | 25 Mn | 26 Fe | 27 Co | 28 Ni | 29 Cu | 30 Zn | 31 Ga | 32 Ge | 33 As | 34 Se | 35 Br | 36 Kr |
| 37 Rb | 38 Sr | 39 Y | 40 Zr | 41 Nb | 42 Mo | 43 Ma | 44 Ru | 45 Rh | 46 Pd | 47 Ag | 48 Cd | 49 In | 50 Sn | 51 Sb | 52 Te | 53 J | 54 X |
| 55 Cs | 56 Ba | 57 * La | 72 Hf | 73 Ta | 74 W | 75 Re | 76 Os | 77 Ir | 78 Pt | 79 Au | 80 Hg | 81 Tl | 82 Pb | 83 Bi | 84 Po | 85 — | 86 Rd |
| 87 — | 88 Ra | 89 Ac | 90 Th | 91 Pa | 92 U | 93 — | 94 — | 95 — | 96 — | | | | | | | | |
| | | * | 58 Ce | 59 Pr | 60 Nd | 61 — | 62 Sm | 63 Eu | 64 Gd | 65 Tb | 66 Dy | 67 Ho | 68 Er | 69 Tu | 70 Yb | 71 Cp | |

Figure 2. The Periodic System as it was presented by Ida Noddack in 1934 [Noddack, 1934, p. 301]. Though there was no official acknowledgement of the existence of masurium and the discovery was not confirmed, Ida Noddack lists Ma as element 43. In the public domain.

[3] Walter and Ida (Tacke) got married in 1926 soon after the official announcement of their discovery of the two new elements.

Van Assche aimed at rehabilitating the work and credit of the Noddack couple. The rehabilitation process, which he initiated, included different authors and channels of communication, from the most public to the most specialised. An example of a public communication can be found in a journalist's article in the *Frankfurter Allgemeine* in 1989 [Zettler, 1989]. Indeed, the word *Rehabilitation* appears in the subtitle: "Warum das Element 43 Technetium heißt. Rehabilitation von W. Noddack und I. Tacke" ("Why element 43 is named technetium. Rehabilitation of W. Noddack and I. Tacke") [Zettler, 1989, p. N1] and is clearly linked to two figures: Walter and Ida Noddack. Otto Berg's name had by that time apparently disappeared from the discovery history. This may have been because he — a chemist working with the Siemens and Halske research laboratory — had only published one or two further papers on the new element 75 that had been named rhenium and nothing on element 43. There are actually very few records on Otto Berg as he was of Jewish origin and had to flee Germany for London, where he died in 1939. Retrospectively, Ida Noddack is better known for her suggestion that the nucleus could be split, made as early as 1934, that is, four years before Otto Hahn (1879–1968), Fritz Strassmann (1902–1980), Lise Meitner (1878–1968), and Otto Robert Frisch (1904–1979) announced the existence of nuclear fission [Van Tiggelen, 2019]. According to Van Assche, the failure of the Noddacks to actually isolate a significant amount of their alleged masurium had undermined their scientific credibility among the larger community, and for that reason, Ida's proposal was simply ignored even when later proven right.

Most of Van Assche's target audience was, however, part of the scientific community itself. Van Assche was caressing the hope that by demonstrating that the Noddacks had indeed *observed* element 43 back in the 1920s, they would receive some retrospective recognition. By "observing", he was referring to the X-ray spectrum. The Berlin team had accompanied their claim to discovery with a series of data, one of them being a photograph of spectral lines typical of element 43, taken with an X-ray spectrograph, in one of their first publications [Berg and Tacke, 1925]. This was actually the only time a spectrum was provided for masurium, and according to Segrè's autobiography, when he asked Walter Noddack to see the original plate in 1937, he was told it had been broken [Segrè, 1993, p. 209]. Van Assche's line of reasoning is as follows: because of the very elusive nature of element 43, the radioelement could not be isolated and produced in sufficient quantities, but some of the minerals investigated by Noddack and his colleagues contained uranium, of which element 43 was a fission product; therefore, the X-ray spectrum published in 1925 could very well have shown the lines expected for that element. The main argumentation thus addressed two questions: (1) whether it was possible that element 43 could have been present and detectable in the sample described by Noddack and his colleagues, and (2) whether they had indeed observed the presence of element 43. The latter part of the discussion focused on the interpretation of the photograph since the original plate was not available and what information could be extracted from it, taking into account present knowledge [Van Assche, 1988].

The hypothesis immediately aroused interest and triggered many responses, some positive and enthusiastic about setting historical records straight, and others questioning the assumptions made to demonstrate that the Berlin team had "seen" element 43. Many radiochemists, however, were not convinced, and the year after, German nuclear scientist Günter Herrmann (1925–2017) followed up with a severe critique of Van Assche's proposition published in the same journal, *Nuclear Physics* [Herrmann, 1989]. According to Herrmann, concentrations of naturally occurring technetium are very low and several orders of magnitude below the detection limit given by the Noddacks and Berg for their technique. Herrmann also underlined that the German team mentioned samples containing masurium that were not uranium, a point also stressed by another radiochemist, the Japanese-American chemist and nuclear scientist Paul Kuroda (1917–2001), who entered the controversy at the same time in the same journal [Kuroda, 1989]. Van Assche was not deterred and teamed up with American spectroscopist John T. Armstrong, who simulated the X-ray spectra that would be expected from Van Assche's initial estimates of the Noddacks' residue compositions

[Armstrong and Van Assche, 1999]. With that last argument, which did not prompt any change of opinions among the community of radiochemists and was entirely based on a computer-based reconstruction of what could have been observed compared to a photograph, one had now drifted far from any actual historical evidence provided by the Berlin-based team.

Van Assche attempted what historians call presentism, i.e., judging the past in the light of present knowledge and techniques. Most nuclear scientists who entered the debate had really thought they were working with historical material and, on the basis of this material, believed they stuck as close as possible to the work and experience of the German team in the 1920s. But this was not the case, and the spectrum reconstructed through simulation was indeed the last stretch of an *ad hoc* argument. However, more than the lack of historisation of pieces of evidence that were circulated and discussed, the whole debate left out something more important: the context of discovery. The Noddacks were looking for specific elements in the frame of a periodic system still based on atomic weight and on the notion that all elements are to be found in nature. In other terms, it would not have made sense for them to look for disintegration products to identify the missing element, which is precisely what the controversy triggered by Van Assche was focusing on. The Noddacks' horizon of expectations was framed by the periodic system before radioactivity and the discovery of neutron, even though they were very much aware of these new scientific developments [Van Tiggelen, 2001; Van Tiggelen and Lykknes, 2019]. By contrast, the debates occurring in the 1980s in the wake of an attempt to revisit the discovery chronology focused the discussion on technical aspects typical of the nuclear physics involved in the production and manipulation of heavy chemical elements. Even if Noddack and his colleagues were to be retrospectively recognised for having observed traces of element 43, they were doing this observation while looking for *eka*-manganese (and also *dvi*-manganese), a manganese homologue to be extracted from ores. They reasoned that its presence would be signalled by properties interpolated in the way Mendeleev organised his own prediction of what would become gallium, scandium and germanium.[4] Van Assche, Herrmann and others were, however, concerned with element 43, or its isotopes eventually produced in a cyclotron in 1937 and since then manufactured for medical technical purposes; Walter and Ida Noddack kept looking for natural element 43 as an *eka*-manganese, isolated through the so-called wet-chemical method late in the 1950s [Van Tiggelen, 2001, p. 241].

## Act II: Reassigning Nipponium

In 1908, a Japanese chemist, Masataka Ogawa (1865–1930), published two notes announcing the discovery of a new element in a thorianite mineral coming from Ceylan [Ogawa, 1908a; Ogawa, 1908b]. The analysis of this material had been proposed to him in 1904 by William Ramsay (1852–1916), in whose laboratory he was then working, before going back to Japan where he resumed the work, extended the investigation to ores available in Japan, and published the first results of his research. In his conclusion, among the 12 properties that demonstrate that the substance is a new element, he mentioned the atomic weight as the second to last in his list:

> 11. The equivalent weight of the element is about 50, this element probably filling the gap between molybdenum and ruthenium and having an atomic weight of about 100. [Ogawa, 1908, p. 10]

According to Ogawa, it was Ramsay who suggested naming the new element nipponium after Ogawa's homeland, which is Nippon in Japanese. Despite an academic career with a heavy teaching load and administrative duties, Ogawa carried the investigation further, but except for a lengthier contribution in 1909 and some reports between 1911 and 1917, all of them in Japanese, he did not publish more on the matter, and nipponium gradually vanished from the international chemical literature [Hisamatsu *et al.*, 2022].

[4] See also Gisela Boeck's contribution to this volume.

At the end of the 20th century, in a paper published in *Radiochimica Acta*, the case of the discovery of nipponium was explored by the Japanese radiochemist H. Kenji Yoshihara (1932–2022), who suggested that the new element Ogawa had isolated and observed could have been what is now called rhenium (element 75) [Yoshihara, 1997]. The story of nipponium bears some similarities with that of masurium: a claimed discovery that was never confirmed. It has indeed caught the attention of scientists and historians alike over the last 25 years, triggering passionate debates, the unearthing of hitherto untapped historical material and the reappraisal of scientific evidence [Kaji, 2003; Yoshihara *et al.*, 2005]. The naming of element 113 as nihonium (an alternative phonation from the Japanese for Japan) in 2016 and the thriving of both science and the history of science in Japan has led to new insights in this intriguing story of nipponium, with the careful analysis of historical material in Japanese and a detailed examination of the published observations [Hisamatsu *et al.*, 2022]. But in contrast with masurium, nipponium is now assigned retrospectively to another element (rhenium), which complicates the story of element 43 and provides a new perspective into the process of retrospective attribution when reassessing previous claims of discovery. In this second part of the chapter, I will draw on the results of the recent research to clarify the nature of the arguments at stake without assessing them and deciding whether or not credit should be given. Instead, the point indeed is to understand how, why, and what at first was claimed to be the element with an approximate atomic weight of 100 became a forerunner for an element with a little less than double its atomic weight.

When Ogawa left London for Tokyo in 1906, he brought with him the thorianite material to further pursue his research on the new substance he had isolated and observed. He also investigated other ores available in Japan containing minerals, such as reinite and molybdenite, which, as he noted in his first publication two years later, he suspected might also have contained the new element. His notes were very detailed: the analytical separation and the determination of the properties (atomic weight and visible spectrum) were successively reported, leading to the conclusion that describes the features of a new element he suggested to call nipponium. Ogawa also mentioned a substance that he set aside and which he believed might also contain another new element [Ogawa, 1908a]. In his second publication, Ogawa reported his work on that latter substance, from which he extracted a new element whose atomic weight was estimated to be 100. "But the number of 100 is that already given to nipponium as its atomic weight," he wrote, "and nipponium differs from the element now under consideration in its oxide not being reducible by hydrogen, and in other respects" [Ogawa, 1908b, p. 9]. He considered the possibility that instead of being bivalent, nipponium might prove trivalent, in which case the atomic weight would be 150, but he left this discussion to after the results on specific heat had been obtained by a colleague. Further examination of fraction samples brought back from London led Ogawa to point at the possibility of yet other elements to be discovered in thorianite [Ogawa, 1908b]. Despite the ambiguity about the atomic weight, periodic table enthusiasts such as Frederick Henry Loring immediately incorporated the new element in their design (Figure 3).[5]

The question of the atomic weight and the constitution of higher oxides were, of course, linked, and the correct placement of the putative element in the periodic system before British chemist Henry Moseley's (1887–1915) atomic number system could only be achieved with both of these questions settled. In contrast with what was the core in the work of Noddack, Tacke and Berg discussed in the first part of this chapter, the periodic system in the early 20$^{th}$ century could not yet fully perform as a decisive prediction tool. After Moseley's law in 1913, predictions were more straightforward; the missing elements could be unequivocally and easily identified through their atomic numbers and their characteristic X-ray spectra. One of the hypotheses to explain Ogawa's public silence on the sci-

[5]Nipponium and the symbol Np also made its way into F. H. Loring's atomic weight tables, one of which has a question mark since it had not yet been confirmed by the International Committee of Atomic Weights. See [Loring, 1909].

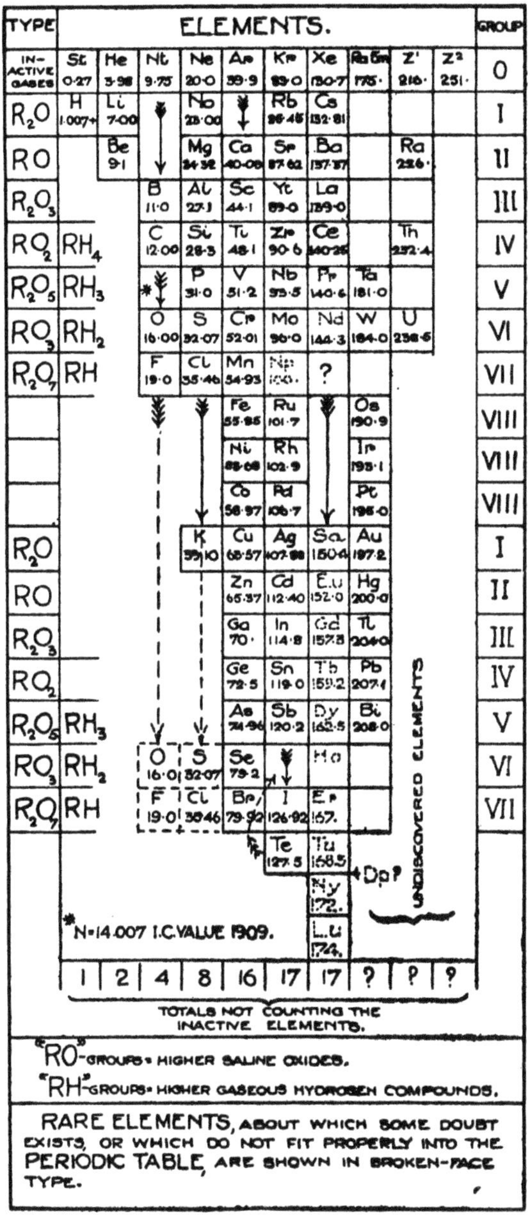


Figure 3. F.H. Loring's table of elements [Loring, 1909, p. 283]. The symbol "Np" for nipponium, placed to the right of Mn, appears in broken-face type, as nipponium is among the elements for which some doubts exist. In the public domain.

entific stage despite evidence that he was still working on nipponium was his constantly postponed hope to submit his samples to X-ray spectrography [Kaji, 2003]. Somewhat tragically, this measure was eventually achieved a few weeks before Ogawa passed away in 1930 and demonstrated that samples from molybdenite ores indeed contained rhenium. By then, the Berlin team Noddack, Tacke and Berg had been acknowledged as discoverers of element 75 (rhenium), while their claim to have discovered element 43 (masurium) remained unsubstantiated, as we have learned in Act I. The 1930 X-ray results from Ogawa's younger colleague were published a couple of years later and were to become the main piece of evidence for a profound reconsideration of the nipponium story from a new perspective: the discovery of another manganese homologue that is *dvi*-manganese instead of *eka*-manganese, which Ogawa had initially claimed to discover.

Taking over the matter at the end of the 20$^{th}$ century, Yoshihara was thus not the first to suggest that instead of technetium, Ogawa might have actually isolated rhenium, but with the increasing interest raised since his first publication about 25 years ago (in 1997), he was the first to pursue the matter over a long period of time and with new historical documents. In particular, he was able to access the papers and relics of Ogawa kept by his family, as well as gather recollections from many colleagues and students, including those who had actually witnessed or participated in Ogawa's research at the end of his life [Yoshihara, 2000 and 2004]. Among the studied artefacts was the photographic plate of the 1930 X-ray spectrum and the related sample, which allowed Yoshihara to proceed with a modern measure that confirmed the presence of rhenium. From that point forward and up until recently, much work has been put into reconsidering the whole of Ogawa's publications and making sense of the wet-chemical analyses, or the visible spectral lines, by reinterpreting nipponium as rhenium [Yoshihara, 1997; Yoshihara, *et al.*, 2002; Hisamatsu *et al.*, 2022].

The present discussions on nipponium entangle two levels, with both a scientific and a historical reevaluation. The latter is problematic as Ogawa's work is assessed against a claim he did not make himself. Here, as was the case with the Noddacks, the horizon of Ogawa's expectations is ignored, even though in this case, the reevaluation of evidence is based on the historical material. Ogawa was looking for element 43 and found traces of element 75, but he is retrospectively assigned as a forerunner for discovering 75, a claim he himself never made — at least in published writing. In contrast with the Noddacks, who were looking for element 43 and would, according to Van Assche, have observed it without being able to isolate it, Ogawa's work is now reassigned to a missing element from the periodic system different from which he had originally identified. The naming of element 113 as nihonium honours both Ogawa's fine work as a chemist and the development of Japanese expertise in radiochemistry [En'yo Ho, 2019; Meija, 2021].

## Act III: Recollecting Forerunners: Polinium, Pelopium, Ilmenium, and Davyum

Getting to other "forerunners" of technetium further back in time, the plot thickens. As potential instances of what would later be named technetium, the English Wikipedia page dedicated to element 43 lists polinium, ilmenium, pelopium, davyum, and lucium[6] before mentioning nipponium, with a date and a "discoverer" for each of them [Wikipedia, n.d.]. These names can also be found in the chemical review literature like [Hackney, 1951] and [Murin *et al.*, 1961], for instance, and in publications for the general and curious public, e.g., [Ball, 2021]. Interestingly, most of the putative elements mentioned in the list above have not attracted a lot of historical analysis despite the potential light these stories can shed on the question of element discoveries. An exception is the encyclopaedic book *The Lost Elements. The Periodic Table's Shadow Side*, co-authored by Marco Fontani, Mariagrazia Costa, and Mary Virginia Orna in

[6] The case of lucium is not investigated in this chapter. For a history of lucium, see [Fontana *et al.*, 2015, pp. 165–170].

2015 [Fontani *et al.*, 2015]. This book focuses precisely on all elements that were eventually not accepted and were, as the authors put it, "lost". The three chemistry historians have provided a detailed and informed account of all elements previously claimed to have been discovered and their fate until their dismissal or fall into oblivion. Their historically documented approach does not aim at retrospectively assessing — or rehabilitating — actors and their claims; their purpose is to tell an informed story on how these claimed discoveries emerged, how they were discussed and criticised, and why these substances were lost to the list of chemical elements. In Act III, I will scrutinise these lost elements as forerunners to further unflesh the process of retroactive attribution. Why and how indeed are dismissed discoveries of elements reinterpreted and reassigned in a hugely different time and context when data and documents are scarce and much less accessible, as in the case of masurium and nipponium?

In an article published in 1828 in *Poggendorff's Annalen*, the German chemist Gottfried Wilhelm Osann (1796–1866) announced the isolation of three new metals he named polinium,[7] pluralium and ruthenium, extracted from platinum ores coming from the Ural region [Osann, 1828a; 1828b]. Soon after, in 1829, Osann had to retract his claim after Swedish chemist (Jöns) Jacob Berzelius (1779–1848), who had initially supported his claim, analysed a sample that he had been sent and in which he could not find any trace of unknown metals [Osann, 1829]. An element, also named ruthenium, was, however, later to be discovered by Carl Klaus (1796–1864) in these exact same ores [Fontani *et al.*, 2015, pp. 73–77]. Apart from the fact that platinaferous material was long considered a potential source of manganese homologues, there is nothing to link polinium to element 43. The same goes with pelopium, which German mineralogist and chemist Heinrich Rose (1795–1864) claimed to have discovered in 1844. He also announced the discovery of another element, niobium, but while niobium was eventually confirmed as a

[7] Initially named "polin"; see [Osann, 1828b, p. 352].

new element, the claim for pelopium was rapidly dismissed by another element hunter, Hans Rudolf Hermann (1805–1879), also a German chemist who made his career in Moscow [Fontani *et al.*, 2015, pp. 46–47].

Hermann and Rose were contenders in other controversies surrounding new metals from the tantalum family as well. Historian of chemistry Sarah Hijmans underlines the continuity of analytical methods to identify (new) metals and how the debates focused on the properties and chemical behaviour of their compounds in their possible purest form [Hijmans, 2022]. Hermann's claim in 1847 to have identified yet another new tantalum metal is rooted in that tradition, which also rests on mineralogical knowledge and practice. Though Hermann provided an atomic weight for a new metal he named ilmenium, this is but one characteristic on which he established his claim to a new element, and this characteristic was by far not the most critical in his view. Along with the kind of ores investigated (among other columbite), it is probably the recalculated atomic weight Hermann provided in 1871 that invited much later for an identification of the "lost element" ilmenium with the present technetium. Indeed, in his last publication, he provided an atomic weight of 654 (much lower than the 786,59 published earlier) for ilmenium, which in old atomic scales with oxygen = 100, would be 104,64, thus not far from the atomic weight for element 43. This *post hoc* identification, made implicitly by listing ilmenium as a forerunner to element 43, however, completely overlooks the analytical tradition that uses many other properties to substantiate the identification and the identity of a new substance and ignores the lengthy results Hermann published on the different characteristic of the compounds — physical, chemical but mainly mineralogical, contrasted with other elements of the tantalum family, mostly niobium, along with a detailed expose on minerals containing these elements.

Published in 1871, Hermann's last contribution on ilmenium does not refer to the periodic system [Hermann, 1871]. This is understandable as the classification proposed in 1869 had not yet gained traction among the chemical community [Kaji *et al.*, 2015]. Conversely, the claims for davyum (in 1877) and lucium (in 1896), also among the "lost" elements [Fontani *et al.*, 2015, pp. 165–170], pointed to a specific place in the periodic system, which can better support their attribution to *eka*-manganese than mere closeness with the present atomic weight of element 43. This indicates that the search for new elements after 1869 could be framed within a system that not only compares a new element with all the others to make sure it was indeed a novelty but also situated it in a sequence in which the chemical and physical properties evolve as a series of related elements are considered. I will concentrate here on the dynamic story of davyum.

In 1877, a Russian chemist, Sergius Kern, sent a short notice to several journals announcing that he had isolated a new metal, which he called davyum (in honour of the English chemist Humphry Davy (1778–1829)), as he was investigating platinum residues that were sent to him from the Ural mines [Kern, 1877a; 1877b]. Working with minute amounts of material after removing several other metals from the platinum group, he determined characteristic behaviours of different compounds and placed the so-called davyum between molybdenum and ruthenium in the fifth period of Mendeleev's system. In his own words:

> The new metal may be supposed to occupy the place between molybdenum and ruthenium in the periodical system of elements proposed by Prof. Demetrius [*sic.*] Mendeleeff (*Annalen d. Chemie und Pharmacie*, Suppl. B, viii., p. 133; *ibid.*, p. 168). Perhaps Davyum is the hypothetical metal in the periodical system having the atomic weight equal to 100. [Kern, 1877a, p. 11]

Kern's first estimate of 100 as the atomic weight for the new element is consistent with this suggestion, but a later determination by a colleague yields a possible atomic weight of 154 and, surprisingly, no mention was made that this was no longer consistent with Mendeleev's predictions for *eka*-manganese [Kern, 1877b]; see also

| Valence. Series. | I. | II. | III. | IV. | V. | VI. | VII. | VIII. | | | I. | II. | III. | IV. | V. | VI. | VII. | VIII. |
|---|---|---|---|---|---|---|---|---|---|---|---|---|---|---|---|---|---|---|
| 1. | .. | .. | .. | .. | .. | .. | .. | .. | .. | .. | .. | .. | H | He | * | (+) | (+) | (?) |
| 2. | Li | Be | B | C | .. | .. | .. | .. | .. | .. | .. | .. | .. | A | N | O | F | (?) |
| 3. | Na | Mg | Al | Si | .. | .. | .. | .. | .. | .. | .. | .. | .. | (+) | P | S | Cl | (?) |
| 4. | K | Ca | Sc | Ti | V | Cr | Mn | Fe | Ni | Co | Cu | Zn | Ga | Ge | As | Se | Br | (?) |
| 5. | Rb | Sr | Y | Zr | Nb | Mo | Da | Ru | Rh | Pd | Ag | Cd | In | Sn | Sb | Te | I | (?) |
| 6. | Cs | Ba | La | Yb | Ta | W | Url | Os | Ir | Pt | Au | Hg | Tl | Pb | Bi | Ast | (+) | (?) |
| 7. | (+) | Ms | (+) | Th | Np | U | (?) | (?) | (?) | (?) | | | | | | | | |

$_4$III.*a* = Sc, Pp. $_4$I.*b* = Cu, Ng. $_6$III.*a* and $_6$IV.*a* = La, Ce, Ny, Py, Sm, Gd, Tb, Ho, Er, Tm, Dc, Yb. . . .

* Unnamed element accompanying helium.

Figure 4. Periodic table by F. Rang. Note the presence of Da (davyum) underneath Mn. Rang includes several "half-forgotten" elements in a table he claims is the "truest and best tabular arrangement of the elements yet produced" [Rang, 1895, p. 201]. In the public domain.

Figure 1. Further publications no longer referred to atomic weight either but to visible spectral lines and characteristic reactions of compounds and insisted on the expected extreme rarity of davyum in nature [Kern, 1878a; 1878b]. In fact, Kern never mentioned the periodic system in his later publications again, giving an indication that at the time, and in his mind, the positioning in the frame of the periodic system was less of a criterion than singular chemical and physical behaviour and spectroscopic lines. In other words, his horizon of expectations had not completely absorbed the new classification system. Some authors, such as F. Rang (see Figure 4), however, identified davyum with *eka*-manganese unequivocally and included the "new" element in their proposed periodic table.

The discovery claim was neither further defended nor retracted until 20 years later, when the American chemistry professor John William Mallet (1832–1912) repeated Kern's procedures and showed that the analytical work and the weight determination of 154 were consistent with his own analysis but that the substance was, in fact, a mixture of rhodium and iridium with traces of iron [Mallet, 1898]. The interest for the substance did not rest so much in the historical straightening of records, though Mallet certainly mentioned that the existence of davyum was not confirmed, but it is the expectations on the properties of the forecasted element that were attracting attention as well as the feeling that there were missing elements in the platinum group:

> We may say, however, that not yet is there any reliable evidence of any new metal between the two series, that is, with an atomic weight of about 150; nor has there been any trace of *eka*-manganese, with its atomic weight of 100, and which some chemists would expect to find resembling the platinum metals in its properties. [Howe, 1900, p. 1020]

By the 1920s, Moseley's reinterpretation of the periodic system in terms of atomic numbers allowed chemists to write of missing (and desired) elements, both in terms of their atomic number (in this case, element 43) and their closest homologue (in this case, *eka*-manganese). Reviewing all the claims of new elements in 1927, British chemist John Gerald Frederick Druce (1894–1950) dismissed the case of davyum [Druce, 1927]. More than 20 years after, however, he came back to the case, and analysing the publications more closely with the British chemist and textbook writer John Albert Newton Friend (1881–1966), he came to the conclusion that the substance manipulated by Kern could, in fact, be *dvi*-manganese. While they went over all the analytical details, the main argument for this assertion was the atomic weight since when "preliminary determination indicated a value of about 154, it was obvious that davyum could not be *eka*-manganese though it might be *dvi*-manganese" [Friend and Druce, 1950, p. 819]. They concluded that

> There is, of course, no doubt of the existence of element 75 in Nature; the point is, does it exist in platinum ores? If it does not, then davyum is definitely ruled out. But if it should at some future date be found to occur in platinum minerals, that would go a long way towards substantiating Kern's claim to have discovered a new element. [Friend and Druce, 1950, p. 819]

As it happens, there are minute traces of rhenium in platinum ores, as Ida and Walter Noddack had already demonstrated [Noddack and Noddack, 1927, pp. 355–356]. Eventually, it seems a long-discarded claim to discovery was revised by Friend and Druce, and the author of this discovery claim, Kern, rehabilitated in his honour so to speak: his observations were considered accurate after all. This reassessment does not take into account the status of Kern, whose scientific work was often criticised [Allen, 1877a; 1877b; Riley, 1878], and the fact that Kern himself did not follow up on his claim after 1878. While Kern did mention that the new metal could have been *eka*-manganese in his first announcement, this also was not at the core of his claim — he never followed up on this aspect, focusing instead on the analytical characterisation of the substance, and most of all, the atomic weight determination was, according to his own comments, still tentative because the amount of substance was so limited. In fact, the reinterpretation of Kern's claim by Friend and Druce shifted from hinting at *eka*-manganese to having described *dvi*-manganese, a move that is nowhere to be found in Kern's publication, even after the new estimate for the atomic weight did not agree with what Mendeleev suggested for *eka*-manganese.

It is even to be noted that Druce himself completely changed his assessment of Kern's claim between 1927 and 1947, as his own horizon of expectations was evolving. In 1927, the discovery of *dvi*-manganese, or element 75, was still disputed, and he did not associate Kern's claim with that element specifically, while in 1947, rhenium, henceforth accepted as element 75, was present in every periodic table across the world and only then was the hypothesis made that Kern might have observed rhenium. To sum up, Kern's context and horizon of expectations were simply ignored, and by reassessing his work, the story unfolds as if an element was there waiting to be discovered, and the issue is to retrospectively identify which element Kern could have observed. The full story of that element, which is no longer element 43 but element 75, then needs to bear witness to this chemist who definitely observed the substance before those now-considered discoverers.

Polinium, ilmenium and pelopium appeared in the literature before Mendeleev proposed the periodic system, while davyum and lucium appeared after. Without having the periodic system as a frame, one might then wonder how the attribution is made retrospectively and on which argument these claimed elements were presented as forerunners to element 43. Obviously, the minerals investigated, the analytical characterisation, and mostly the atomic weight all come into account but are retrospectively interpreted differently. This retrospective look on forerunners, such as davyum, provides an opportunity to understand what motivates the writing of an elemental discovery as a biographical narrative of hide-and-seek and also how scientists are making sense of past errors, reconciling the community with their predecessors and long-passed colleagues, and, therefore, illustrating that science is a cumulative endeavour across space and time.

## Conclusion

These few examples from the history of the manganese homologues, which do not even cover all instances of element 43's forelife, demonstrate how the process and the history of discovery can be approached from different points of view in time and space, providing successive narratives. If one thinks in terms of biographies, it looks as though element 43 knew successive lives, or to be more precise, incarnations, and that the process of discovery is often encapsulated in a very simple assertion: this element was discovered here at that time by this-and-that chemist. Not only is there a need to unfold the process of discovery and the surrounding controversies, but, as historian of chemistry Frank James argued when reviewing the case of thallium and the contemporary discussion about what constitutes the discovery of a chemical element (detection, verification or isolation), one "cannot simply consider only the science ... but we also have to examine and define the context" in which the discovery is developing to understand the motivation of the individuals partaking in it [James, 1984, p. 65]. To some extent, the same could be said about the cases of masurium, nipponium, davyum, and ilmenium when presented as former apparitions of element 43 in the chemical literature of the past.

As Act I reveals, Noddack, Tacke and Berg would be credited for having observed a long-lived isotope of technetium, whereas they were looking for natural 43 and desperately trying to isolate it through wet-chemical methods. The same analytical virtuosity was Ogawa's way to reach for nipponium but was later reported as the discovery of *dvi*-manganese despite Ogawa's own published claims, as explained in Act II. In Act III, the retrospection goes back even further in time — Hermann and Kern could well have observed a new element and it could well have been what is now called 43 (or 75 for that matter), but the association of ilmenium and davyum with *eka*-manganese cannot stand unequivocally since Hermann and Kern were not thinking in the frame of our present understanding of what chemical elements were to be in a periodic system reinterpreted by X-rays and how to unequivocally identify them.

The fact is historically clear: These claims to discovery were not recognised in their own times. Even though they might indeed have observed something new, the actors themselves continued their quest (or never withdrew their claims), which means the goal was not yet considered as achieved within their own horizon of expectations. It is only later that when scientists and historians alike try to find the "biography" of the element, these unfamiliar stories of "forerunners" emerge and try to rectify the accuracy of a short and dry "birth certificate" by extending the quest back in time to more actors and events. However, these new stories, while they aim at doing justice to past endeavours in the history of chemical sciences, fail to question the encapsulation of the process and the perspective of the narrator(s). Furthermore, they are making sense of evidence retrospectively with modern tools based on sometimes very thin historical evidence and documents. The successive lives of element 43 thus demonstrate that the horizon of expectations is embedded mostly unconsciously in the narrator's point of view in two ways: first, by ignoring the experience and expectations of the actors and second, by overlooking their own expectations in assessing scientists' work and discoveries. The first bias seems the most remarkable because whenever the discussion or the controversy actually arises, more often than not, the aim is actually to rehabilitate, reevaluate or reconsider. But, precisely, re-evaluation is a process rooted in the horizon of expectations ... of the narrators.

Discovery is far from elemental, even when it comes to elements. In fact, narratives of element discovery, which would be typically thought to be more straightforward than any other discovery stories precisely because of the elemental nature of its object, turn out to be good starting points to unveil all the complexity of the discovery process. Not only does it raise the question of how elements are defined and identified at a certain point in the time of the development of science, but it also shows how the assessment of that process is reworked through successive narratives, creating new biographies for a seemingly simple scientific object. It also uncovers the formidable emotional weight of being a discoverer, which explains why scientists retrospectively attempt to revisit, reevaluate and rehabilitate, going as far as reassigning the work of their predecessors. In the case of element 43, this happened more than once, which makes it a perfect case to dig into the matter more deeply and, thereby, lift the veil on the underlying assumptions and practices of crediting and acknowledging discoveries and on the importance of discovery in the scientific fabric and the identity of the scientific community.

## Acknowledgements

The author thanks Annette Lykknes and two reviewers for their most valuable feedback on the manuscript.

## References

Allen, A. H. (1877a). To the Editor of the Chemical News. *Chemical News*, 20 July 1877, p. 33.

Allen, A. H. (1877b). To the Editor of the Chemical News. *Chemical News*, 21 December 1877, p. 274.

Armstrong, J. T. and Van Assche, P. (1999). The disputed discovery of Element 43: A reexamination of an elegant early use of wavelength-dispersive X-ray microanalysis. *Journal of Research of the National Institute of Standards and Technology*, 104, p. 599.

Ball, P. (2021). *The Elements: A Visual History of Their Discovery* (Chicago: The University of Chicago Press).

Berg, O. and Tacke, I. (1925). Zwei neue Elemente der Mangangruppe: Röntgenspektroskopischer Teil, *Sitzungsberichte der physikalisch-mathematischen Klasse vom 11. Juni 1925* (Königlich Preussische Akademie der Wissenschaften zu Berlin), pp. 405–409.

Druce, J. G. F. (1927). Elements whose Existence has been announced, but which are not recognised. *Science Progress in the Twentieth Century*, **21**(83), pp. 479–491.

En'yo, H. (2019). History of nihonium. *Pure and Applied Chemistry*, **91**, pp. 1949–1958. http://doi.org/10.1515/pac-2019-0810

Fontani, M., Costa, M. and Orna, M. V. (2015). *The Lost Elements: The Periodic Table's Shadow Side* (Oxford: Oxford University Press).

Hackney, J. C. (1951). Technetium — Element 43. *Journal of Chemical Education*, **28**(4), pp. 186–190.

Hermann, R. (1871). Fortgesetzte Untersuchungen über die Verbindungen von Ilmenium und Niobium, sowie über die Zusammensetzung der Niobmineralien. *Journal für praktische Chemie*, **112**, pp. 178–210.

Herrmann, G. (1989). Technetium or masurium — a comment on the history of element 43. *Nuclear Physics*, **A505**, pp. 352–360.

Hijmans, S. (2022). The tantalum metals (1801–1866): Nineteenth-century analytical chemistry and the identification of chemical elements. *Ambix*, **69**(4), pp. 399–419.

Hisamatsu, Y., Egashira, K. and Maeno, Y. (2022). Ogawa's nipponium and its re-assignment to rhenium. *Foundations of Chemistry*, **24**, pp. 15–57. https://doi.org/10.1007/s10698-021-09410-x

Howe, J. L. (1900). The eight group of the periodic system and some of its problems. *Science, New Series*, **11**(287), pp. 1012–1020.

James F. (1984). Of 'medals and muddles' the context of the discovery of thallium: William Crookes's early spectro-chemical work. *Notes and Records of the Royal Society of London*, **39**(1), pp. 65–90.

Kaji, M. (2003). Ogawa's discovery of a new element "nipponium": The emergence of modern chemistry research in Japan and its social background. *Historia Scientiarum*, **12**(3), pp. 238–241.

Kaji, M. (2015). Chemical classification and the response to the periodic law of elements in Japan in the nineteenth and early twentieth centuries. In *Early Responses to the Periodic System*, eds. Kaji, M., Kragh, H. and Palló, G. (Oxford: Oxford University Press).

Kaji, M., Kragh H. and Palló, G., eds. (2015). *Early Responses to the Periodic System* (Oxford: Oxford University Press).

Kern S. (1877b). On some new researches on the metal davyum. *Chemical News*, **36**, p. 1144.

Kern, S. (1877a). On the new metal davyum (Preliminary notice). *Chemical* News, **36**, p. 4.

Kern, S. (1878a). On the spectrum of the metal davyum. *Chemical News*, **37**, p. 33.

Kern, S. (1878b). Some remarks on the metal davyum. *Chemical News*, **37**, p. 65.

Kuroda, P. K. (1989). A Note on the discovery of technetium. *Nuclear Physics*, **A503**, pp. 178–182.

Loring, F. H. (1909). The atomic wieghts as mathematical functions. *Chemical News*, 10 December 1909, pp. 281–286.

Mallet, J. W. (1898). On the claims of davyum to recognition as a chemical element. *American Chemical Journal*, **20**, p. 776.

Meija, J. (2021). Politics and the periodic table. *Nature Chemistry*, **13**, pp. 814–816.

Mendeleev, D. I. (1871). Die periodische Gesetzmäßigkeit der chemischen Elemente. *Annalen der Chemie und Pharmacie*, suppl. vol. 8, pp. 133–229.

Murin, A. N., Nefedov, V. D., Ryukhin, Yu. A. and Toropova, M. A. (1961). Technetium — Element 43. *Russian Chemical Reviews*, **30**(2), pp. 107–115.

Noddack, I. (1934). Über das Element 93. *Angewandte Chemie*, **47**, pp. 653–656.

Noddack, I. and Noddack, W. (1927). Das rhenium. *Ergebnisse der exakten naturwissenschaften*, **6**, pp. 333–373.

Noddack, W. and Tacke, I. (1925). Zwei neue elemente der mangangruppe: Chemischer teil, *Sitzungsberichte der physikalisch-mathematischen Klasse* vom 11. Juni 1925 (Königlich Preussische Akademie der Wissenschaften zu Berlin), pp. 405–409.

Ogawa, M. (1908a). Preliminary note on a new element in thorianite. *Chemical News*, **98**, pp. 249–251.

Ogawa, M. (1908b). Preliminary note on a new element allied to molybdenum. *Chemical News*, **98**, pp. 261–264.

Osann, G. (1828a). Fortsetzung der Untersuchung des Platins vom Ural. *Poggendorffs Annalen der Physik und Chemie*, **13**, pp. 283–297.

Osann, G. (1828b). Fortsetzung der Untersuchung des Platins vom Ural. *Poggendorffs Annalen der Physik und Chemie*, **14**, pp. 329–357.

Osann, G. (1829). Berichtigung, meine Untersuchung des uralschen Platins betreffend. *Poggendorffs Annalen der Physik und Chemie*, **15**, p. 158.

Perrier, C. and Segrè, E. (1947). Technetium: The element of atomic number 43. *Nature,* **159**, p. 24.

Rang, [P. J.] F. (1895). The period. table. *Chemical News*, 25 October 1895, pp. 200–201.

Riley, E. (1878). To the Editor of the Chemical News. *Chemical News*, 31 May 1878, pp. 226–227.

Segrè, E. (1993). *A Mind Always in Motion. The Autobiography of Emilio Segrè* (University of California Press).

Van Assche, P. (1988). The ignored discovery of the element Z = 43. *Nuclear Physics* **A480**, pp. 205–214.

Van Tiggelen, B and Lykknes, A. (2019). A tale of resilience: The periodic table after radioactivity and the discovery of the neutron. *Centaurus*, **61**(4), pp. 287–298. https://doi.org/10.1111/1600-0498.12245

Van Tiggelen, B. (2001). The discovery of new elements and the boundary between physics and chemistry in the 1920s and 1930s: The case of elements 43 and 75. In *Chemical Sciences in the 20$^{th}$ Century: Bridging Boundaries*, ed. Reinhardt, C. (Wiley-VCH, Weinheim).

Van Tiggelen, B. (2019). Ida Noddack, the eka-manganeses and nuclear fission. In *Women in their Element: Selected Women's Contributions to the Periodic System,* eds. Lykknes, A. and Van Tiggelen, B. (World Scientific, Singapore), pp. 209–224.

Wikipedia. (n.d). Technetium, https://en.wikipedia.org/wiki/Technetium (retrieved March 2023).

Yoshihara K., Kobayashi, T. and Kaji, M. (2005). Ogawa family and their 'nipponium' research: Successful separation of element 75 before its discovery by Noddacks. *Historia Scientiarum*, **15**, pp. 177–190.

Yoshihara, K. (1997). Nipponium, the element ascribable to rhenium from the modern chemical viewpoint. *Radiochimica Acta*, 77, pp. 9–13.

Yoshihara, K. (2000). Ogawa's discovery of nipponium and its reevaluation. *Historia Scientiarum*, **9**, pp. 257–269.

Yoshihara, K. (2004). Discovery of a new element 'nipponium': Reevaluation of pioneering works of Masataka Ogawa and his son Eijiro Ogawa. *Spectrochimica Acta*, **B59**, pp. 1305–1310.

Yoshihara, K., Nakai I., Terada Y. and Kaji M. (2002). Nipponium studies by Masataka and Eijiro Ogawa. *Journal of the Japanese Society for the History of Chemistry*, **29**, pp. 209–221.

Zettler, H. (1989). Masurium — ein Name, den keiner mehr nennt. *Mittwoch 22. February 1989*, **45**, p. N1.

## Selected Bibliography

Fontani, M., Costa, M. and Orna, M. V. (2015). *The Lost Elements: The Periodic Table's Shadow Side* (Oxford: Oxford University Press).

Hijmans, S. (2022). The tantalum metals (1801–1866): Nineteenth-century analytical chemistry and the identification of chemical elements, *Ambix*, **69**(4), pp. 399–419.

Hisamatsu, Y., Egashira, K. and Maeno, Y. (2022). Ogawa's nipponium and its re-assignment to rhenium. *Foundations of Chemistry*, **24**, pp. 15–57. https://doi.org/10.1007/s10698-021-09410-x

Kaji, M., Kragh H. and Palló, G., eds. (2015). *Early Responses to the Periodic System* (Oxford: Oxford University Press).

Van Tiggelen, B. (2001). The discovery of new elements and the boundary between physics and chemistry in the 1920s and 1930s: The case of elements 43 and 75. In *Chemical Sciences in the 20$^{th}$ Century: Bridging Boundaries*, ed. Reinhardt, C. (Wiley-VCH, Weinheim).

https://doi.org/10.1142/9789819813858_0008

# NOBELIUM, THE QUESTIONABLE ELEMENT

Ann E. Robinson

*Widener Library, Harvard University, Cambridge, MA 02138, USA*
*ann_robinson@harvard.edu*

The last years of the 1930s marked a change in the discovery of new chemical elements. In 1937, element 43, technetium, was the first element to be created in a laboratory. Two years later, element 87, francium, was the last element to be discovered in nature [Weeks and Leicester, 1968].[1] Between 1940 and 1955, 11 new elements were discovered. Atoms of all of these elements were synthesised and created in a laboratory using a particle accelerator.[2] Within the accelerator, ions of isotopes of elements were accelerated to a high energy level.[3] These ions were then flung towards a target covered

[1]Some elements first synthesised in laboratories have later been found in nature. Trace amounts of technetium, plutonium, promethium, astatine, and neptunium have been found in ores containing uranium as a result of radioactive decay. See also Brigitte Van Tiggelen's contributions to this volume.

[2]Particle accelerators come in two basic forms: cyclotrons and linear accelerators. In a cyclotron, ions run in a circular track while gaining energy on each half-turn before being directed towards a target; in a linear accelerator, oscillators and amplifiers are used to increase the energy of ions. Ions are atoms stripped of one or more electrons.

[3]Isotopes are atoms of the same element that have different atomic masses due to the different number of neutrons in the atom's nucleus.

with an ultrathin layer of isotopes of a heavier element, bombarding it and causing nuclei to fuse and particles to fly away. The hoped-for result of this bombardment was the creation of atoms of an entirely new element. As the atomic numbers of the new elements increased, fewer and fewer atoms of them were created. The discovery of mendelevium, element 101, was based on the creation of only 17 atoms of one isotope of the new element [Ghiorso *et al.*, 1955]. Peter Armbruster (b. 1931), a physicist at the *Gesellschaft für Schwerionenforschung* (GSI) in Darmstadt, Germany, referred to 1955–1965 as the years of apprenticeship in which scientists were learning how to develop experiments and read their results [Armbruster, 1985]. The synthesis of elements heavier than element 101 was very much a frontier in nuclear science.

The official recognition and naming of these new elements were also entering a new frontier.[4] Before the use of particle accelerators, new elements were usually found in minerals or in gases. During the 19th and early 20th centuries, there were generally accepted standards for announcing such a discovery but no official criteria. These general criteria were explained by one chemist in 1912 as being, "first, to establish some positive proofs of existence of a new substance (in the first line the atomic weight, the spectrum, and some reactions); second, to publish a short preliminary notice in any of the numerous scientific papers" [Estreicher, 1912, p. 119].[5]

Atomic weight[6] was considered to be the determining characteristic of an element, hence the importance of establishing the atomic

---

[4] The second half of the 20th century was a period when new bureaucratic processes arose in many areas of chemistry. For example, the question of what substances were considered to be chemicals became of interest in regard to toxicology and environmental control; see [Hepler-Smith, 2019].

[5] The criteria are discussed in Giunta's and Kragh's contributions. See also Lykknes' contribution to this volume.

[6] The terms atomic weight and atomic mass are often used interchangeably when referring to historical times, although they are different entities. Today, the atomic weight of an element is the weighted mean value of the relative isotopic masses of the naturally occurring isotopes of that element, while atomic mass refers to the sum of the masses of the atom's constituents, namely, protons, neutrons and electrons

weight of a newly discovered element. This changed in the 1920s, as chemists and physicists learned more about the structure of the atom, and atomic weight was no longer viewed as the most important characteristic of an element [Kragh, 2000]. In 1923, the International Union of Pure and Applied Chemistry (IUPAC), the standards-making body for the international chemical community, declared that an element was "defined by its atomic number", that is, the number of protons in the nucleus of an atom [ICCE, 1923, p. 868].[7] Determining atomic number became an important addition to the informal criteria that were in use.

It is not easy, or perhaps even possible, to determine the atomic number, atomic weight, spectra, or chemical reactions when you synthesise a mere handful of atoms of a radioactive element that decays within minutes if not seconds or milliseconds. If these previously accepted criteria no longer function, then how is it possible to determine if a new element has been discovered? And if more than one laboratory discovers atoms of a new element but different isotopes of that element, who is the one that discovered the element, and who gets to name it? Decades of controversy over the discovery of new elements, as well as advances in experimental techniques and in the understanding of atoms, were required before the creation of the first official set of criteria for determining if a new chemical element had been discovered. The discovery and naming of element 102 forced chemists and physicists to begin to grapple with the questions that led to these new criteria.

***

All of the new elements synthesised since 1940 were done so either at the University of California, Berkeley, and its associated

---

(the mass of the electron is, however, negligible). Before the structure of the atom was known, atomic weight was determined by painstakingly comparing and calculating the weights of elements in compounds. For more on the history of atomic weight and atomic mass, see [Coplen and Holden, 2011] and [IUPAC, 1992].

[7] Today, the IUPAC is in charge of such things as determining atomic weights, defining terms, and defining the nomenclature of chemical substances.

Radiation Laboratory or with the involvement of its scientists.[8] Credit for the discovery of elements 99 and 100, both of which were discovered in the debris of the 1952 Ivy Mike bomb test, was shared between Berkeley and the Los Alamos National Laboratory in New Mexico, as well as the Argonne National Laboratory (ANL) near Chicago, Illinois. Paul R. Fields (1919–1998), a chemist at the ANL, was one of those who worked on these elements.

In 1955, Fields and his colleague Arnold M. Friedman (1926–1985) resolved to search for element 102 "independently", without the involvement of the Berkeley scientists [Holl, 1997, p. 180]. The heaviest element that was available in large quantity at the ANL was curium. Thus, the plan was to bombard targets of a mixture of curium atoms — $^{244}$Cm, $^{245}$Cm and $^{246}$Cm — with carbon atoms of $^{13}$C. Synthesising new elements involves some mathematics. When you bombard a heavier atom with a lighter one, it needs to create an atom with whatever atomic number you are seeking to make. You need to take into account not only the protons in the atom's nucleus but the electrons and neutrons, as well. Fields and Friedman determined that the following nuclear reactions were likely to produce atoms of element 102:

$$^{244}\mathrm{Cm} + {}^{13}\mathrm{C} \rightarrow {}^{253}102 + 4n$$
$$^{244}\mathrm{Cm} + {}^{13}\mathrm{C} \rightarrow {}^{251}102 + 6n$$

If they bombarded a target of $^{244}$Cm with atoms of $^{13}$C, it should produce atoms of element 102 with atomic mass 253 and four neutrons or atoms of element 102 with atomic mass 251 and six neutrons.

The difficulty in synthesising element 102 "independently" was two-fold. While the ANL had the curium isotopes, they did not have a sufficient quantity of $^{13}$C. Argonne also did not have an accelerator capable of reaching the high energy levels needed to

[8] The Radiation Lab was renamed the Lawrence Berkeley Laboratory in 1959 after the death of Ernest O. Lawrence (1901–1958), an inventor of the cyclotron.

carry out the bombardment. Fortuitously, in 1956, John Milsted (1917–1984) was visiting the ANL from the Atomic Energy Research Establishment in Harwell, England. Milsted was going to be at the Nobel Institute for Physics in Stockholm, Sweden, for a short period as part of an exchange program, and he would have access to their powerful cyclotron.[9] Harwell would also be able to supply the needed $^{13}C$. The hunt for element 102 would thus be an international collaboration.

Fields was able to join Milsted in Stockholm for only two weeks. It wasn't until the day before Fields had to return to Argonne that the bombardment runs yielded 20 atoms of what appeared to be element 102. The "most probable" isotopes created were $^{253}102$ and $^{251}102$, with a half-life of approximately 10 minutes. It was predicted that even-numbered isotopes would have a half-life of approximately 10 seconds and were thus not observable. There were several reasons for the use of words such as "probable" and "approximately". It was suspected that several of the targets had impurities. It was also believed that there was radiation damage in some of the targets, which could have caused a reduction in the number of atoms created [Fields *et al.*, 1957, p. 1461]. Depending on the level of radiation damage, not only can targets be contaminated, but so too can equipment, and it can cause faulty data. For example, an accident at the Berkeley heavy ion linear accelerator (HILAC) in July 1959 destroyed the $^{244}Cm$ targets being used in element 102 experiments and contaminated the area with curium "so heavily that it was unusable" and resulted in part of the HILAC being rebuilt [Hoffman *et al.*, 2000, p. 244].[10]

Another difficulty lay in the short half-life of some of the isotopes of element 102. All radioactive elements begin to decay after a certain period of time, leaving half of the total amount of atoms after that time. The time it takes to transform half of the atoms is specific

[9] In 1954, users of the cyclotron at the Nobel Institute believed they had discovered element 100. However, due to the secrecy surrounding atomic bomb tests, they were not aware that this element had already been discovered in the Ivy Mike test debris [Chapman, 2019, pp. 97–98].

[10] For an account of the accident and its aftermath, see [Garden and Dailey, 1959].

to each nuclide or species of an atom[11] and is known as the half-life. As they decay, the atom breaks apart into a combination of particles and other elements. Uranium atoms of $^{238}U$, for example, have a half-life of approximately 4.5 billion years, after which half of the initial sample decays into thorium atoms of $^{234}Th$ and alpha particles. After a long series of decays, what was $^{238}U$ ultimately became $^{206}Pb$, a non-radioactive, stable isotope of lead. While the atoms of element 102 that were created in Stockholm had half-lives of about 10 minutes, other isotopes of element 102 were predicted to have half-lives of mere seconds. An article on the discovery in *Time* magazine entitled "Chemists, Run!" described how, after the bombardment, chemists had to remove the target from the cyclotron, dismantle it, and then run to a laboratory on a different floor where alpha particles from the disintegrating atoms could be measured.[12] "With practice," the article noted, they were able to "cut the total elapsed time to 1 1/2 minutes" [Time, 1957, p. 40]. One and a half minutes may not sound like much time, but it is enough to lose precious atoms from an already small sample of element 102 to decay.

Seventeen atoms of element 101 had been enough for a discovery claim only two years before, so 20 atoms was enough for the Stockholm team to publish the results of the bombardments and confidently state the discovery of element 102. In their *Physical Review* paper, the team wrote, "We suggest the name nobelium, symbol No, for the new element in recognition of Alfred Nobel's support of scientific research and after the institute where the work was done" [Fields *et al.*, 1957, p. 1461]. The discovery was big news, not only in the scientific community but also among the general public. Newspaper headlines highlighted the international

[11]The term nuclide refers to a specific atomic species or type of atom, whereas the term isotope more strictly refers to a species of a specific element. The term nuclide was first introduced in 1947 but did not gain widespread use until the 1960s, so many papers until that time used isotope to refer to a species of atom. For more on the creation of the term nuclide, see [Kohman, 1947].

[12]Chapman describes the cyclotron laboratory at the Nobel Institute as "equipment made by hand on a shoestring budget" as they did not have "Berkeley's vast pockets". Nevertheless, "they astonished the world" [Chapman, 2019, p. 114].

Figure 1: Paul R. Fields with Swedish Consul General Goesta Oldenburg pointing to the space on the periodic table for the newly discovered element 102, July 1957 (United Press/Newspaper Enterprise Association; photo in author's personal collection).

nature of the discovery, as did the accompanying photographs (Figure 1). The story made the front page of the *New York Times*, followed several days later by a more in-depth article describing the experiments, while the headline of the *Associated Press* article, which found its way into newspapers from the United States to India, blared, "Tenth Atomic Age Element Found" [Johnston, 1957; Laurence, 1957; AP, 1957].

## The Rush to Naming

When element 102 was discovered in 1957, the official rules for naming elements were being created by the Commission on Nomenclature of Inorganic Chemistry (CNIC) of the IUPAC. From 1900, a new element was considered to be official when the International Committee on Atomic Weights (ICAW), in 1920 to become part of the newly founded IUPAC, added it to the international table of atomic weights with the name and symbol given to it by its discoverer(s), along with its atomic weight [Urbain, 1924]. This practice continued until just after World War II when the ICAW first shared and then ceded the official naming of new elements to the CNIC [Holden, 2004].

The CNIC had begun drafting official rules for inorganic nomenclature in the 1930s, but the process was interrupted by World War II. They took up the work again at the first post-war meeting in 1947 and ten years later were finalising the rules for publication. As part of the process, the CNIC harmonised the names of several existing elements, particularly the ones that were known by more than one name, such as elements 74 (wolfram and tungsten) and 41 (columbium and niobium) [Robinson, 2019]. This resulted in some basic rules regarding element naming. The suffix *-ium* was considered to be distinctive of chemical elements — symbols should be in two letters, names and symbols should be considerate of as many languages as possible, and in what became Rule 1.12, element names should carry "no implication regarding priority of discovery" [IUPAC, 1957, p. 6].

The IUPAC was meeting in Paris when it was publicly announced that element 102 had been discovered. The CNIC meeting minutes have a palpable excitement about the prospect of a new element. The minutes of their 11 July morning meeting record, "The Chairman told the meeting of a newspaper report concerning the alleged Swedish discovery of element 102, to which the name 'Nobelium' was stated to have been given." Arne Ölander (1902–1984), a professor at the University of Stockholm, was asked to contact the Nobel Institute for Physics "at once ... to seek confirmation of the report in order that the proposed name might come before the Commission during the present series of meetings".[13] It was seen as important that the discovery be confirmed quickly. The CNIC

[13] Othmer Library of Chemical History, Science History Institute, Philadelphia, PA. International Union of Pure and Applied Chemistry Archive. Box 41. I.U.P.A.C. Inorganic Nomenclature Commission. Paris Meeting. Thursday, 11 July 1957. Morning Session.

generally met only every other year, and a delay would mean a possible wait of at least two years for the name to become official. As the rules of inorganic nomenclature were to be sent to the publisher soon after the meeting, element 102 would not be included on the list of elements if there was a delay.

Two days later, Ölander reported that he had received a telegram from Manne Siegbahn (1886–1978), the director of the Nobel Institute for Physics, where the discovery took place, confirming the discovery of element 102. The CNIC then agreed to include element 102, nobelium (with the symbol "No"), in the list of element names in the rules.[14] In general, the submission of element names and symbols came directly from the discoverer(s) in the form of a letter to the IUPAC. The meeting minutes note that as the notification of discovery came from Siegbahn, who was not directly involved in the work, "the status of the discovery would be regularised if one of the team of discoverers wrote directly to the Commission".[15] Accordingly, the CNIC chairman, Alexander Silverman (1881–1962), wrote to Fields in August. Perhaps Fields was no longer feeling as confident about the discovery as he had been, as his reply was "very informative", detailing the experiments and who was involved. Silverman noted, "It is not the function of the Commission on Inorganic Nomenclature, I.U.P.A.C., to determine details of discovery claim [*sic*], but we do want to file these details for future reference in case any question of any kind is raised."[16]

If there were questions, it would not be the first time there was a dispute over a newly discovered element. In the past, the ICAW

---

[14] Othmer Library of Chemical History, Science History Institute, Philadelphia, PA. International Union of Pure and Applied Chemistry Archive. Box 41. I.U.P.A.C. Inorganic Nomenclature Commission. Paris Meeting. Saturday, 13 July 1957. Afternoon Session.

[15] Othmer Library of Chemical History, Science History Institute, Philadelphia, PA. International Union of Pure and Applied Chemistry Archive. Box 41. I.U.P.A.C. Inorganic Nomenclature Commission, Paris Meeting. Wednesday, 17 July 1957.

[16] Othmer Library of Chemical History, Science History Institute, Philadelphia, PA. International Union of Pure and Applied Chemistry Archive. Box 42. Letter from Alexander Silverman to Paul Fields, 20 September 1957.

had avoided having to deal with such issues. They had typically waited to add an element to the international table until a "measurable amount" of an element had been separated and its atomic weight determined. "The process usually took a considerable number of years, by which time any potential problems with the discovery of the element were usually apparent in the scientific literature" [Holden, 2008]. Radium, for example, was not added to the international table of atomic weights until five years after its discovery was first reported [Clarke, 1903].

In 1957, element 102 was officially named nobelium (with the symbol "No"), and it was included in the first official rules of inorganic nomenclature [IUPAC, 1957]. Nobelium was also included on the table of radioactive elements that accompanied the international table of atomic weights that year [Wichers, 1958]. The ICAW had decided in 1949 that as "true atomic weight is a property of an element as it occurs in nature ... the mass number of the most stable known isotope" would be used instead for "artificial" elements [Wichers, 1950, p. 1431]. However, nobelium was such a recent discovery that its most stable isotope was still unknown. Thus, it appeared on the table with only a name and symbol. This name and symbol soon began to appear on new revisions of periodic tables as well (Figure 2).

## Attempts to Confirm the Discovery

The general rule in science is that before something can be considered to have been truly discovered, the results of the original experiments have to be replicated and confirmed, preferably by someone other than the original discoverer [NAP, 2019]. After the element 102 bombardments were run at the Nobel Institute for Physics in Stockholm, the cyclotron was rebuilt for other types of experiments, and it was not possible to reproduce the original experimental environment [Holl, 1997]. This meant that there were only two laboratories in the world that had particle accelerators capable of running such experiments: the Lawrence Berkeley Laboratory

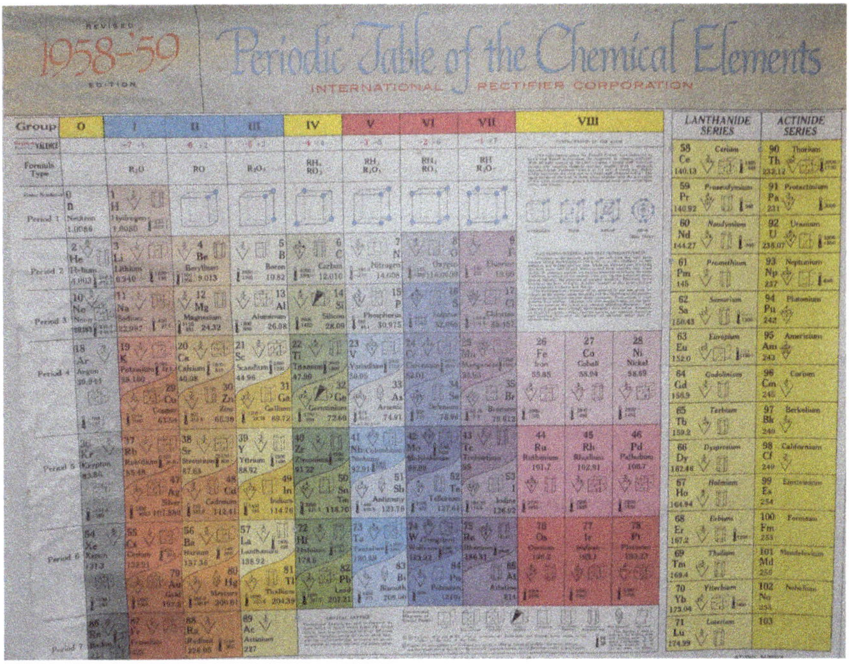


Figure 2: The Periodic Table of the Chemical Elements, included on the International Rectifier Corporation's 1958–59 advertising calendar (International Rectifier Corporation, El Segundo, CA; table in the author's personal collection).

(LBL), perched on the hill above the university campus in Berkeley, California, and the newly opened Joint Institute for Nuclear Research (JINR), located in Dubna, Russia, outside of Moscow. Scientists at both institutions had read the paper in *Physical Review* reporting the discovery of element 102 and set out to confirm it.

Experiments at LBL commenced in the fall of 1957. Albert Ghiorso (1915–2010) led the team undertaking the hunt for the new element. The Berkeley team created targets of curium in a similar mix of isotopes as those used in Stockholm. They also used $^{13}C$ atoms to bombard the curium atoms in the Berkeley HILAC. The opening sentence of the paper detailing their work relayed their findings: "In many score of experiments conducted in various ways over a period of many months we find that we are unable to confirm the element-102 discovery work of Fields *et al.* reported in 1957" [Ghiorso *et al.*, 1958a, p. 17]. They detected isotopes of other elements but no isotopes of nobelium that matched those reported. The Berkeley group noted that the experiments carried out in Stockholm were "extremely difficult" to perform. Also, the energies used were unlikely to result in an isotope of nobelium that was heavy enough to have a half-life of 10 minutes.

Soviet scientists concurred with this evaluation. In a review of the work done in the year after the Stockholm experiments, chemist Vitalii Iosifovich Gol'danskii (1923–2001) expressed the same reservations as the Berkeley group. He also noted that as the atomic number becomes higher, the instability of an atom grows. Therefore, isotopes of elements with an atomic number greater than 100 are more likely to have half-lives measured in seconds rather than minutes. Thus, even before the first work on element 102 began, "it was clear that the limit of applicability of any chemical methods of identifying new elements was already near, and that new methods must be found" [Gol'danskii, 1959, p. 183]. This threw doubt on the chemical work done in Stockholm as a means of identifying nobelium.

In their attempts to replicate the results of the Stockholm experiments, the Soviet group at Dubna, led by Georgii Nikolaevich Flerov (1912–1990), used a different experimental setup in their cyclotron. They bombarded target atoms of plutonium with ions of oxygen — more precisely, $^{239}$Pu and $^{241}$Pu with ions of $^{16}$O — at a much higher energy than used in the Stockholm accelerator. Several atoms of what could be nobelium were created at a similar energy level to those reported in Stockholm; however, they had half-lives in the region of fractions of a second rather than approximately 10 minutes [Flerov *et al.*, 1958]. As Gol'danskii noted, this work "does not signify a confirmation of the Stockholm data" [Gol'danskii 1959, p. 185].

Fields and his co-discoverers addressed the failure of both Berkeley and Dubna to replicate their results. As noted, the Stockholm cyclotron had been rebuilt, so they were unable to attempt to experimentally reproduce their results. Instead, they reexamined the data from their original experiments. They emphasised the differences between their experimental setup and the one used at Berkeley, which could account for the different results. Berkeley used a similar mixture of curium isotopes in the targets, but these were arranged in a stack, and the Stockholm group utilised a single thick target. The Berkeley target arrangement resulted in what they believed was an "improved" recoil technique, which was more efficient in capturing the atoms of element 102 when they were created and leaving less time for decay before measurement [Fields *et al.*, 1959, pp. 225–226].[17]

The Stockholm group argued that the mass assignment they made to nobelium, $^{251}$No or $^{253}$No, was based on information available at the time; however, new information made it likely that these mass assignments were incorrect and "must be modified". They now thought it possible that atoms of $^{255}$No could have been synthesised, which would explain the half-life of about 10 minutes. Aside from this mass reassignment, their "re-examination of experimental data has not led to any new conclusions regarding the interpretation of our results". They believed further experimental studies were necessary, and until then, "judgement on the discovery of element 102 should be reserved" [Fields *et al.*, 1959, p. 228].

## The Re-Discovery of Element 102

In some respect, the Dubna group agreed with Fields and his co-discoverers that judgement should be reserved. In the same article in which the Dubna group stated they were unable to confirm the Stockholm results, they noted that they had observed particles that were "in all probability associated with the decay of isotopes of element 102". Further experiments were already being planned "using an improved method" [Flerov *et al.*, 1958, p. 548]. Experiments at

---

[17] A drawing of the Berkeley apparatus can be found in [Ghiorso *et al.*, 1958b]. This is the same equipment that was involved in the accident described in [Garden and Dailey, 1959].

Dubna to both confirm the mass assignments of the isotopes of nobelium and create more isotopes continued for several years, with updates being communicated throughout the 1960s.

In 1968, Flerov published a review of the experiments related to nobelium carried out in Stockholm, Berkeley and Dubna. He concluded that "errors were committed in most of the work during 1957–1961", including work at Dubna. It was at Dubna, however, where isotope $^{252}No$ was synthesised, information regarding the alpha decay of six different isotopes were first obtained, and the chemical behaviour of nobelium first studied [Flerov, 1968, p. 14]. In other words, in the view of the Soviet team, nobelium was discovered at Dubna, not in Stockholm.

Although Berkeley also continued to carry out experiments related to element 102 after 1957, they were not so hesitant to claim discovery. The paper in *Physical Review Letters* immediately following the one in which they stated they were unable to confirm the Stockholm results was a discovery announcement of their own. After describing their experiments, the Berkeley group stated there was "no doubt that we have identified element 102 in these experiments". They had synthesised isotope '$^{254}102$'. The paper also stated, "We are deferring the naming of element 102 to a later date" [Ghiorso *et al.*, 1958b, p. 20].

This statement is interesting given the fact that element 102 had been officially named nobelium by the IUPAC in 1957. It should be noted that neither Berkeley nor Dubna referred to element 102 as nobelium in their papers. Both groups referred to it exclusively as element 102 and used 102 in place of a symbol (e.g., $^{252}102$ rather than $^{252}No$). Neither recognised that the Stockholm group were the true discoverers of element 102 and thus refused to recognise the name and symbol chosen by them. As Dubna group members Yuri Ts. Oganessian (b. 1933) and Ivo Zvára (b. 1934) explained, "Names, as known, are suggested by the authors of the papers that have been recognized to possess the priority in discovery" [Oganessian and Zvara, 1993, p. 240].

## Can You Rename an Element?

Although the Berkeley group quickly claimed the discovery of element 102, they did not immediately write to the CNIC with their suggestion for a name and symbol. Rather, the matter was brought to the attention of the CNIC by Edward Wichers (1892–1984), an American member of the IUPAC Commission on Atomic Weights. In a 1963 letter to E. J. Crane (1889–1966), who was the editor of *Chemical Abstracts* as well as a member of the CNIC, Wichers noted that "a number of people" had questioned the continued use of nobelium in the international table of atomic weights given the multiple discovery claims.[18]

As the IUPAC was meeting in England not long after, Crane immediately wrote to Glenn T. Seaborg (1912–1999), a Nobel laureate at Berkeley who had been involved in all of the element discoveries beginning with element 93, at LBL. Seaborg responded with what he referred to as a "description" of the discovery of element 102 at Berkeley, but which also included a lengthy refutation of the results from both Stockholm and Dubna. He opined, "It is unfortunate that the Commissions on Atomic Weights and Inorganic Nomenclature acted so promptly in 1957, and apparently without competent technical advice."[19] Although the Berkeley group continued to refer to element 102 as such, Seaborg informed Crane that "we don't have any recommendation for a name for element 102 at this time".[20]

Shortly after the IUPAC meeting in 1963, Crane reported to Seaborg that the CNIC "freely admit that they were apparently a bit hasty in approving the name nobelium, even though the reported

[18] Manuscript Division, Library of Congress, Washington, DC. Glenn T. Seaborg Papers. Box 843. Letter from Edward Wichers to E. J. Crane, 7 June 1963.

[19] Manuscript Division, Library of Congress, Washington, DC. Glenn T. Seaborg Papers. Box 843. Letter from Glenn T. Seaborg to E. J. Crane, 15 June 1963.

[20] Manuscript Division, Library of Congress, Washington, DC. Glenn T. Seaborg Papers. Box 843. Letter from Glenn T. Seaborg to E. J. Crane, 22 June 1963.

discovery came from three reputable institutions of three different countries". He wrote that they were also willing to reconsider, although they were not yet ready to make a change.[21] The meeting minutes do not reflect this but rather state that it was suggested that Seaborg be asked whether or not he could "accept the name 'nobelium' which is now so widely used".[22] Ghiorso and Seaborg ultimately agreed in 1967 to accept nobelium, a "magnanimous offer" that Kai A. Jensen (1908–1992), the chair of the CNIC, hoped would "put an end to the disagreeable uncertainty caused by our premature approval of this name".[23]

The "disagreeable uncertainty", as they referred to it, however, was far from over. Not long after the Berkeley group agreed to retain the name "nobelium", the Dubna group also published a claim to the discovery of element 102. Following the usual procedure, they wrote to the CNIC in March 1968. They believed that the name "nobelium" "given by the authors of the wrong work should be cancelled". Further, they proposed the name "joliotium" and symbol "Jl" in honour of the French physicist and Nobel laureate Frédéric Joliot-Curie (1900–1958), who was notably a communist.[24]

In a letter to British chemist Ronald S. Nyholm (1917–1971), the CNIC chair, Jensen, questioned whether the question of the discovery of element 102 would ever be settled. The name "nobelium" was already in use, but that would "not prevent the Russians from

---

[21] Manuscript Division, Library of Congress, Washington, DC. Glenn T. Seaborg Papers. Box 843. Letter from E. J. Crane to Glenn T. Seaborg, 13 August 1963.
[22] Othmer Library of Chemical History, Science History Institute, Philadelphia, PA. International Union of Pure and Applied Chemistry Archive. Box 41. Meeting of the Commission on the Nomenclature of Inorganic Chemistry, Brighton, 27 June to 4 July 1963.
[23] Manuscript Division, Library of Congress, Washington, DC. Glenn T. Seaborg Papers. Box 843. Letter from K. A. Jensen to Glenn T. Seaborg and Albert Ghiorso, 10 April 1967.
[24] Othmer Library of Chemical History, Science History Institute, Philadelphia, PA. Addenda to the Records of Union of Pure and Applied Chemistry. Box 202. Letter from G. N. Flerov *et al.* to K. A. Jensen, 15 March 1968.

quarrelling with the Americans" over priority.[25] The CNIC as a whole seemed to agree. The minutes of their 1968 meeting stated that they "had no special competence to judge" these priority disputes. Further, they reiterated Rule 1.12, which stated that name selection carried no implication regarding the priority of discovery.[26]

This decision was communicated in letters to both Flerov and Seaborg after the CNIC meeting. The Berkeley group seemed to accept the decision without question. The Dubna group, on the other hand, did not. They infrequently raised it in letters to the IUPAC about the discoveries of other elements. But perhaps most telling is the way element 102 appeared on Soviet periodic tables into the 1980s (Figure 3). The symbol for element 102, "No", is given in parentheses, indicating that the name is in use but not official. The symbols of other elements that were in dispute appeared in the same fashion.

## Criteria for Discovering a New Element

Following the CNIC's decision in 1968 not to rename element 102, scientists were left in a curious situation. This element had an official name, "nobelium", but it was also referred to as "joliotium" and as element 102. It was not an ideal situation and not what the CNIC preferred. Certainly, such a state was not what was expected from a standard-making organisation, particularly as they had spent several years eliminating variant names for the already known elements [Robinson, 2019]. However, the CNIC believed that they were not the body to adjudicate discovery claims, nor were they qualified to do so. The unofficial criteria of determining atomic number, atomic

---

[25] Othmer Library of Chemical History, Science History Institute, Philadelphia, PA. Addenda to the Records of Union of Pure and Applied Chemistry. Box 202. Letter from K. A. Jensen to R. S. Nyholm, 29 May 1968.

[26] Othmer Library of Chemical History, Science History Institute, Philadelphia, PA. International Union of Pure and Applied Chemistry Archive. Box 48. Meeting of the Commission on the Nomenclature of Inorganic Chemistry, Hellerup, Copenhagen, Denmark, 26th August to 31st August 1968.

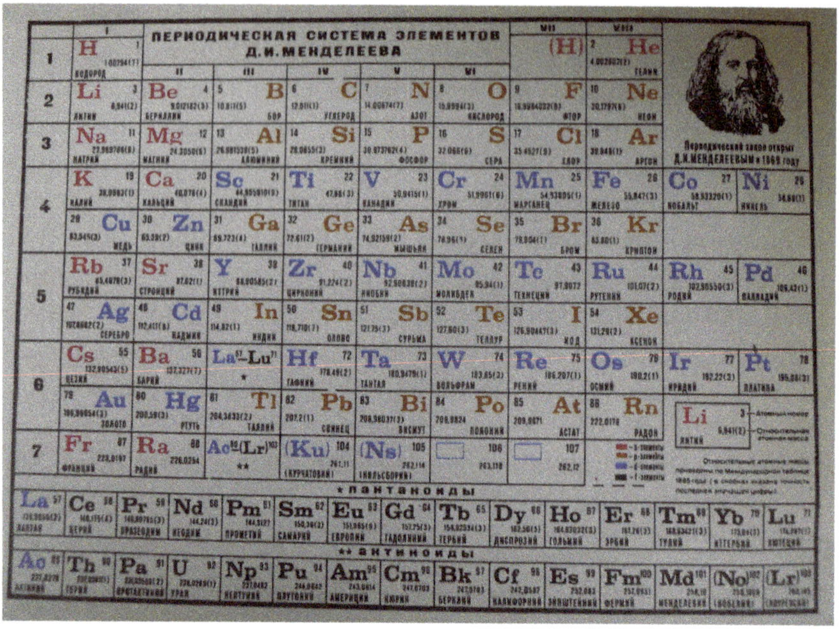


Figure 3: A 1988 Soviet periodic table showing the symbol for element 102, "No", in parenthesis, indicating its name and symbol are not officially recognised (Khimia, Moscow, Russia; table in the author's personal collection).

weight, spectra, and several reactions to show that a new element had been discovered were no longer sufficient. The atomic nuclei of these new heavy elements had half-lives in the range of mere seconds, making it difficult to determine their characteristics and properties. Thus, it was necessary for a new way to be found to answer the question: How do you prove that a new element has been discovered?

Both Dubna and Berkeley had ideas about the criteria that needed to be met to prove discovery. These sets of criteria differed enough as to not be reconcilable. In the 1970s, the IUPAC created a group with the assistance of the International Union of Pure and Applied Physics (IUPAP), known as the Joint Neutral Group (JNG), in order to solve the differences between the Berkeley and Dubna groups. However, the JNG failed to do anything. Subsequently, the IUPAP created a new group with the IUPAC. Formed in the mid-1980s, the Transfermium Working Group (TWG) was explicitly charged with creating a set of criteria for determining if a new element had been discovered.[27]

It took the TWG more than two years and involved meetings with the teams at Berkeley and Dubna and at the GSI in Darmstadt, Germany, which had begun searching for new elements in the 1970s. The work was not straightforward. In their first report, they stated:

> The TWG has also come to the conclusion that it is not feasible to specify criteria, or combinations of criteria, that, in the words of its Terms of Reference 'must be satisfied ...' in order to achieve recognition of the existence of a new chemical element and that would cover all cases.
>
> [Wapstra, 1991, p. 882]

Further, there were "very few properties indeed" that were sufficient unto themselves to establish that a new element had been discovered. The only "uncontentious" example they could offer was "the characteristic X-ray spectrum, unambiguously determined". Rather, "identification must rely upon combinations of properties that will vary from case to case and that cannot usefully be exhaustively codified as a set of criteria" [Wapstra, 1991, p. 883].

The TWG divided the many different criteria into groups.[28] Characterisation properties were used to establish the chemical and/or physical properties of the new element. Among these types of properties were the energy of the bombarding particles used in the experiment, the cross-section (to describe the probability that a nuclear reaction will take place), and the kind of decay (alpha, beta, gamma, or spontaneous fission). Assignment properties were used to show that the characterisation properties were those of a new element. These included such things as cross bombardments (the comparison of probabilities that a specific isotope will be created),

---

[27] For more on the JNG and the TWG, see [Robinson, 2021].

[28] A full list of these criteria and a discussion of how they could be used to demonstrate that a new element had been discovered is detailed in sections III and IV of [Wapstra, 1991].

mass separation (using a mass spectrometer to separate and analyse the products of bombardment), X-ray spectrum (showing the energy of the atoms), and chemical methods, such as ion exchange and gas chromatography. Any combination of characterisation and assignment properties could be used to determine the atomic number of the new element.

The report defined the discovery of a new element as "the experimental demonstration ... of the existence of a nuclide with atomic number *Z* not identified before, existing for at least $10^{-14}$ s". This half-life was the minimum amount of time an atomic nucleus could exist.[29] The TWG also affirmed that "Confirmation demands reproducibility..." and they advocated that "no new element should be recognized officially until the data upon which the claim is based have been reproduced, preferably in another laboratory and preferably by a different technique" [Wapstra, 1991, p. 883].

The TWG then applied their criteria to the elements that were under review — elements 101–109. With regard to element 102, they stated, "It must be considered highly unlikely than an isotope of element 102 was seen in [the Stockholm experiments]" [Wilkinson *et al.*, 1993, p. 1766]. After reviewing 20 papers published by the Berkeley and Dubna groups, as well as others, the TWG concluded that the experiments run at Dubna in 1965 were the first to provide "conclusive evidence that element 102 had been produced" [Wilkinson *et al.*, 1993, p. 1769].

After the publication of this second report, the Dubna group wrote to the IUPAC. Citing the TWG's conclusion regarding element 102, they stated, "We propose that element 102 be named 'Flerovium', after the deceased Professor G. N. Flerov, world famous authority in heavy element research and one of the pioneers of heavy ion physics."[30] Instead, the CNIC chose to accept "the 'status quo'

---

[29] Despite this statement from the TWG, there is no consensus as to the amount of time necessary for an atomic nucleus to exist; see [Kragh, 2018, p. 66].

[30] Othmer Library of Chemical History, Science History Institute, Philadelphia, PA. Addenda to the Records of Union of Pure and Applied Chemistry. Box 203. Letter from V. Shchegolev to IUPAC Secretariat, 7 April 1994.

even though it recognized the conclusion of the TWG that an error had been made in the initial report on the discovery of element 102 (Nobelium)" [CNIC, 1994, p. 2421]. Element 102 had been named "nobelium" (with the symbol "No") in 1957 and would retain that name and symbol.[31] From that moment on, nobelium was no longer a questionable element.

***

Nobelium was not the first synthesised element to be named but it was the first that encountered problems. The CNIC officially named it based solely on a report from the director of the institution where the experiments had been carried out rather than waiting for the discoverers to communicate the discovery directly. No scientific paper by Fields or any member of the Stockholm group detailing the experiments leading to the discovery had yet been published. There had been no replication of experiments or other confirmation. There had not been time for potential problems to become apparent in the scientific literature. The naming of element 102 had been, as the CNIC later acknowledged, rather hasty.

There were reasons for this hastiness. As noted, the CNIC did not meet every year, so it was probable that element 102 would remain officially nameless for at least two years. Although other elements had remained without names for longer, the CNIC was caught up in the excitement of the discovery of a new chemical element occurring as they met. They were in the final stages of preparing the first official rules for inorganic nomenclature. This had been a herculean task, which had taken many years, and they wanted it to be as up-to-date as possible. The CNIC also did not anticipate any problems with regard to the discovery. After all, there had not been any problems with the naming of elements 94–101, which were also synthesised elements.

[31] Flerovium, however, got a second chance in 2012, when element 114 was officially named "flerovium", with the symbol "Fl".

However, nobelium did have disputed discovery claims. Even after Stockholm's claim was discredited by both Berkeley and Dubna, there continued to be competing priority claims and calls to rename element 102. This is due in part to the fact that the two labs making the claims were based in the United States and in the Soviet Union at a time when the Cold War was in a period of crises. National honour was at stake, particularly as these laboratories were state-funded. Discovery and naming were seen as "an acknowledgement of [the discoverers'] intellectual property as well as of the efforts and expenses of the laboratory where the discovery has been made" [Oganessian and Zvara, 1993, p. 240].

Many elements had been known by more than one name for decades, if not centuries before the CNIC officially harmonised the names after World War II. But unlike these elements, the official naming of element 102 was done before the discovery was replicated and accepted by the scientific community. The name remained even after the initial discovery was called into question. This rush to naming the element revealed tensions with the process for naming new chemical elements. The CNIC was composed of inorganic chemists who readily admitted that they did not understand the complexities of synthesising heavy elements. Thus, their emphasis was on being responsible solely for naming and not for adjudicating discovery claims. It also highlighted a tension between scientists who believed they had the right to name their discovery and an international standards organisation that believed in rules for naming that allowed for the removal of the link between discoverers and what they discovered.

Nobelium represented an early step into a new frontier of heavy element research. It was the beginning of what Armbruster had called the age of apprenticeship. Mistakes were made, data were misinterpreted, and new techniques were being created and refined. Ultimately, official criteria were developed by a joint IUPAP–IUPAC group to answer the question: What constitutes discovery in the new era of synthesisation of chemical elements? New official rules for the naming of new chemical elements were developed by the IUPAC, which retained the connection between discovery and naming, allowing discoverers to name the elements they discovered. While this process is not always smooth, it has worked, and the periodic table now holds 118 named elements. The age of apprenticeship is thus over.

## References

AP. (1957). Tenth atomic age element found: International research team's work, *The Times of India*, 10 July 1947, p. 11.

Armbruster, P. (1985). On the production of heavy elements by cold fusion: The elements 106 to 109, *Annual Reviews in Nuclear and Particle Science*, **35**, pp. 135–194.

Chapman, K. (2019). *Superheavy: Making and Breaking the Periodic Table* (Bloomsbury Sigma, London, UK).

Clarke, F. W. (1903). Report of the International Committee on Atomic Weights, *Journal of the American Chemical Society*, **25**, pp. 1–5.

CNIC. (1994). Names and symbols of transfermium elements (IUPAC recommendations 1994), *Pure and Applied Chemistry*, **66**, pp. 2419–2421.

Coplen, T. B. and Holden, N. E. (2011). Atomic weights — No longer constants of nature, *Chemistry International*, **33**(2), pp. 10–15.

Estreicher, T. (1912). Canadium, *Chemical News*, **105**, p. 119.

Fields, P. R., Friedman, A. M., Milsted, J., Atterling, H., Forsling, W., Holm, L. W. and Åström, B. (1957). Production of the new element 102, *Physical Review*, **107**, pp. 1460–1462.

Fields, P. R., Friedman, A. M., Milsted, J., Atterling, H., Forsling, W., Holm, L. W. and Åström, B. (1959). On the production of element 102, *Arkiv för Fysik*, **15**, pp. 225–228.

Flerov, G. N., Polikanov, S. M., Karamian, A. S., Pasiuk, A. S., Parfanovich, D. M., Tarantin, N. I., Karnaukhov, V. A., Druin, V. A., Volkov, V. V., Semchinova, A. M., Oganesian, Iu. Ts., Khalizev, V. I. and Khlebnikov, G. I. (1958). Experiments to produce element 102, *Soviet Physics Doklady*, **3**, pp. 546–548.

Flerov, G. N. (1968). Synthesis and investigation of element 102 (Review), *Soviet Atomic Energy*, **24**, pp. 3–15.

Garden, N. B. and Dailey, C. (1959). High-level spill at the HILAC [UCRL-8919], Lawrence Radiation Laboratory, Berkeley, California, USA.

Ghiorso, A., Harvey, B. G., Choppin, G. R., Thompson, S. G. and Seaborg, G. T. (1955). New element mendelevium, atomic number 101, *Physical Review*, **98**, pp. 1518–1519.

Ghiorso, A., Sikkeland, T., Walton, J. R. and Seaborg, G. T. (1958a). Attempts to confirm the existence of the 10-minute isotope of 102, *Physical Review Letters*, **1**, pp. 17–18.

Ghiorso, A., Sikkeland, T., Walton, J. R. and Seaborg, G. T. (1958b). Element no. 102, *Physical Review Letters*, **1**, pp. 18–21.

Gol'danskiĭ, V. I. (1959). News on elements no. 102, *Soviet Physics Uspekhi*, **67**, pp. 183–186.

Hepler-Smith, E. (2019). Molecular bureaucracy: Toxicological information and environmental protection, *Environmental History*, **24**, pp. 534–560.

Hoffman, D. C., Ghiorso, A. and Seaborg, G. T. (2000). *The Transuranium People: The Inside Story* (Imperial College Press, Singapore).

Holden, N. E. (2004). Atomic weights and the international committee — A historical review, *Chemistry International*, **26**(1), pp. 4–7.

Holden, N. E. (2008). Element 74, the Wolfram versus Tungsten controversy [BNL-81324-2008-CP, Brookhaven National Laboratory, Upton, NY, USA.

Holl, J. M. (1997). *Argonne National Laboratory, 1946–96* (University of Illinois Press, Urbana, IL, USA).

ICCE. (1923). Report of the International Committee on Chemical Elements, *Journal of the American Chemical Society*, **45**, pp. 867–874.

IUPAC. (1957). *Nomenclature of Inorganic Chemistry* (Butterworths, London, UK).

IUPAC. (1992). 'Atomic Weight' — The name, its history, definition, and units, *Pure & Applied Chemistry*, **64**, pp. 1535–1543.

Johnston, R. J. H. (1957). Three-nation team creates an element, labels it Nobelium, *New York Times*, 10 July 1957, pp. 1, 19.

Kohman, T. P. (1947). Proposed new word: Nuclide, *American Journal of Physics*, **15**, pp. 356-357.

Kragh, H. (2018). *From Transuranic to Superheavy Elements: A Story of Dispute and Creation* (Springer, Dordrecht, Netherlands).

Kragh, H. (2000). Conceptual changes in chemistry: The notion of a chemical element, ca. 1900–1925, *Studies in History and Philosophy of Modern Physics*, **31**, pp. 435–450.

Laurence, W. L. (1957). Element no. 102 is added to nature's table by the cooperation of three nations, *New York Times*, 14 July 1957, p. 143.

NAP. (2019). *Reproducibility and Replicability in Science* (The National Academies Press, Washington, D.C.).

Oganessian, Yu. Ts. and Zvara, I. (1993). Response to the Report of the Transfermium Working Group, *Progress in Particle and Nuclear Physics*, **31**, pp. 239–240.

Robinson, A. E. (2019). Order From confusion: International chemical standardization and the elements, 1947–1990, *Substantia*, **3**(2, Suppl. 4), pp. 83–99.

Robinson, A. E. (2021). Attempting neutrality: Disciplinary and national politics in a Cold War scientific controversy, *Centaurus*, **63**, pp. 84–102.

Time. (1957). Chemists, Run! *Time*, 22 July 1957, p. 40.

Urbain, G. (1924). Twenty-five years of research on the yttrium earths, *Chemical Reviews*, **1**, pp. 143–185.

Wapstra, A. H. (1991). Criteria that must be satisfied for the discovery of a new chemical element to be recognized, *Pure and Applied Chemistry*, **63**, pp. 879–886.

Weeks, M. E. and Leicester, H. M. (1968). *Discovery of the Elements*, 7$^{th}$ ed. (Journal of Chemical Education, Easton, PA, USA).

Wichers, E. (1950). Report of the Committee on Atomic Weights of the American Chemical Society, *Journal of the American Chemical Society*, **72**, pp. 1431–1432.

Wichers, E. (1958). Report on atomic weights for 1956–1957, *Journal of the American Chemical Society*, **80**, pp. 4121–4124.

Wilkinson, D. H., Wapstra, A. H., Ulehla, I., Barber, R. C., Greenwood, N. N., Hrynkiewicz, A., Jeannin, Y. P., Lefort, M. and Sakai, M. (1993). Discovery of the transfermium elements, *Pure and Applied Chemistry*, **65**, pp. 1757–1814.

## Selected Bibliography

Chapman, K. (2019). *Superheavy: Making and Breaking the Periodic Table* (Bloomsbury Sigman, London, UK).

Hofmann, S. (2002). *On Beyond Uranium: Journey to the End of the Periodic Table* (Taylor & Francis, London, UK).

Kragh, H. (2000). Conceptual changes in chemistry: The notion of a chemical element, ca. 1900–1925, *Studies in History and Philosophy of Modern Physics*, **31**, pp. 435–450.

Kragh, H. (2018). *From Transuranic to Superheavy Elements: A Story of Dispute and Creation* (Springer, Dordrecht, Netherlands).

Münzenberg, G. (2018). *SuperHeavy Elements: Searching for the End of the Periodic Table* (Manipal University Press, Manipal, India).

https://doi.org/10.1142/9789819813858_0009

# THE HEAVY HYDROGEN ISOTOPES: ANTICIPATIONS AND DISCOVERIES

Helge Kragh

*Niels Bohr Institute, University of Copenhagen, 2100 Copenhagen, Denmark.*
*helge.kragh@nbi.ku.dk*

## Introduction

The hydrogen gas, or "inflammable air", discovered by Henry Cavendish (1731–1810) in 1766, is (in practice) a mixture of two isotopes, which was demonstrated in 1932 with the discovery of deuterium, the mass-2 isotope that makes up about 0.02% of ordinary hydrogen. A few years later, the two hydrogen isotopes were extended with a third one, the radioactive tritium of mass 3. Although tritium exists in nature, its abundance in hydrogen is close to zero, namely, about $10^{-18}$%, and its presence is mainly due to nuclear reactions in the upper atmosphere caused by cosmic rays. Here, a fast neutron may react with nuclei of atmospheric nitrogen according to ${}^{14}_{7}\mathrm{N} + {}^{1}_{0}n \rightarrow {}^{12}_{6}\mathrm{C} + {}^{3}_{1}\mathrm{H}$. Deuterium (D) and tritium (T) are the only ones among the numerous isotopes of different elements that are still granted their own names and chemical symbols. The decision of IUPAC (the International Union of Pure and Applied Chemistry) to disallow isotope names except those of hydrogen indicates the special significance of the two light isotopes [Connolly and Damhus, 2005].

Just as the chemical elements from hydrogen (no. 1) to oganesson (no. 118) once were discovered, so were the numerous isotopes, somehow and by someone. Although discoveries are complex, they can be, in many cases, assigned a discoverer or a team of discoverers. The histories of how the isotopes were discovered are less well known, but in a few cases, they are as interesting and instructive as those of the elements figuring in the periodic table. Most but not all of the usual discovery criteria applied to the elements are also, with some modification, valid for the discovery of isotopes. In this chapter, I start with providing an overview of criteria usually associated with discovery in general. I then examine the historical developments that led to the discoveries of deuterium and tritium and also, as a by-product, the mass-3 helium isotope. Moreover, I include a brief section on the neutron, which, in the 1920s, when it was still a hypothetical particle, was sometimes discussed together with the mass-2 and mass-3 hydrogen isotopes. The time frame is approximately from 1913 to 1939.

As to nomenclature, the names deuterium and tritium only date from the early 1930s. H-2 and H-3 and $H^2$ and $H^3$ were also often used in the past, while today, $^2H$ and $^3H$, are the favoured notations, which I will use hereafter. The historical actors involved in the discovery processes were chemists as well as physicists, with some being physical chemists and others nuclear physicists. Much of the work was interdisciplinary. As far as the discovery of tritium is concerned, it largely took place in laboratories devoted to nuclear physics and with little or no participation of traditionally laboratory-trained chemists.

## On Discoveries and Predictions

The often-used criteria associated with a scientific discovery, whatever its kind, might be summarised in three minimum requirements. A discovery must be (1) novel, (2) presented in public, and (3) supported by a majority of the relevant scientific community.[1] None of these criteria are epistemic, which implies that a claim can be accepted as a discovery at least temporarily even though it is later recognised to be false. Whether concerned with elements, isotopes, or something else, the second criterion is a *sine qua non*. A discovery claim must be publicly available, either in the form of a publication or in a well-publicised lecture or the like. To qualify as a discovery, it is not enough that it is known privately or only to a few persons, such as will be exemplified in this essay. It goes without saying that the third criterion presupposes the second one and also, for that matter, that the stated criteria are far from universally accepted.

The notion of discovery is not an either-or category. It also covers the cases where a discovery claim is accepted by a significant fraction of the scientific community but rejected by another and perhaps equally significant fraction. An example of such a "half-discovery" may be the discovery of triatomic hydrogen, $H_3$, to be discussed below. Moreover, both with respect to elements and isotopes, there are cases in which a work reported at a certain time is only understood as a proper discovery in retrospect after it has been complemented or justified by a later work. This kind of dual discovery is related to but is not the same as what is usually called a rediscovery. Again, we shall meet one such case below, namely in connection with the discovery of tritium. Elsewhere, I offer a fuller discussion of the discovery concept in relation to elements [Kragh, 2019].

While some discoveries are serendipitous or more or less accidental, others are, to some extent, derived from theoretical assumptions preceding the actual discovery. The latter is often said to be predicted discoveries, a well-known example being Dmitrii Ivanovich Mendeleev's (1834–1907) so-called prediction of the elements gallium, scandium and germanium.[2] However, the term "prediction" is tricky as it covers a broad range and is often used indiscriminately.

[1] Edmond Rancke-Madsen, for example, proposed a similar set of rules in 1975. See [Rancke-Madsen, 1975, pp. 301–302].

[2] See also Gisela Boeck's contribution to this volume.

Mendeleev's predictions or grounded expectations of new elements are of a different kind than the rigorous theory-derived predictions known from physics and astronomy. Mendeleev's were analogical and not very sharp, and his periodic system did not rely crucially on them. By contrast, Niels Bohr's (1879–1962) prediction of the so-called isotope effect to be discussed in the following section was deductively based on a definite theory of atomic structure. Had it turned out that there was no isotope effect, Bohr's atomic theory would have been falsified [Kragh, 2012b].

In many cases, the term prediction is simply a misnomer, which would better be replaced by a more appropriate term such as suggestion, speculation, hypothesis, or expectation. Thus, when scientists around 1920 suggested the existence of isotopic particles such as $^{2}H$, $^{3}H$ and $^{3}He$, they merely speculated that these entities might exist. The speculation or guesswork was not supported by either experiment or theory, the only argument being that they were simple proton-electron composites and, therefore, conceivable entities. To speak of such "why-not arguments", as we may call them, as predictions are, at best, an unfortunate terminology.

## Early Suggestions of Mass-3 Isotopes

Given that deuterium ($^{2}H$) is stable and relatively abundant, whereas tritium ($^{3}H$) is unstable with a half-life of 12.3 years and only exists naturally as trace amounts, one might expect that the possibility of the former isotope was discussed before the latter, but curiously, the history of $^{3}H$ precedes that of $^{2}H$. While the possible existence of $^{3}H$ was suggested as early as 1913 — the same year and yet before the word "isotope" had been coined — the possibility of $^{2}H$ only turned up in scientific literature in 1919.

In experiments carried out in Cambridge at about 1912 with positive ions, Joseph John Thomson (1856–1940) identified a component with mass 3, which he called $X_3$, and suggested that it was due to a hitherto unknown triatomic hydrogen molecule of the form

$H_3$ [Kragh, 2012a]. Shortly after having published his groundbreaking paper on atomic theory in the summer of 1913, Bohr attended a meeting of the British Association for the Advancement of Science in Birmingham in which Thomson gave a talk on his $H_3$ hypothesis. In the discussion session that followed the presentation, Bohr suggested as an alternative that $X_3$ might be a mass-3 isotope of hydrogen rather than a triatomic molecule. The suggestion was not accepted in the scientific community, but it was known to at least some physicists and chemists. In a letter to Ernest Rutherford (1871–1937) on 14 October 1913, the chemist George Hevesy (1885–1966) wrote:

> Bohr — in his usual modest way — suggested the possibility of $X_3$ being an H-atom with one central charge, but having a three times so heavy nucleus than Hydrogen. ... Bohr's suggestion is that $X_3$ is possibly a chemically non-separable element from Hydrogen. Of course it is not very probable, but still a very interesting suggestion, which should not be quickly dismissed. [Eve, 1939, p. 224]

After having returned to Copenhagen, Bohr realised that the hypothesis might be tested spectroscopically by means of what is known as the "isotope effect". According to Bohr's new atomic theory, the hydrogen spectral lines should depend slightly on the nuclear mass by the factor $(1 + m/M)$, where $m/M$ is the ratio between the mass of the electron and that of the nucleus. He consequently expected the hypothetical $^3$H isotope, if it existed, to turn up as a very small but measurable shift in wavelength, which he calculated to $\Delta\lambda/\lambda = 3.6 \times 10^{-4}$. Bohr looked for this isotope effect, but he found nothing, and presumably for this reason, he did not refer to the hypothesis in any of his publications. He briefly mentioned the spectroscopic isotope effect in a discussion at the British Association meeting in 1915, but at the time without referring to the heavy hydrogen isotope [Kragh, 2012b]. Thus, Bohr's idea of the $^3$H isotope was largely unknown to contemporary scientists. The predicted spectral shift for the $^2$H isotope would be approximately 1.5 times larger than the one of the $^3$H isotope and thus easier to detect, but because Bohr's thoughts were focused on Thomson's $H_3$ molecule, he did not contemplate the possibility of the simpler $^2H$ isotope.

The $^3H$ isotope only turned up in the scientific literature in 1920, first in investigations of the isotopic composition of hydrochloric acid (HCl) made by William Harkins (1873–1951), a chemist at the University of Chicago. Harkins [1920a] not only found evidence for the chlorine isotopes of mass 35 and 37, such as Francis Aston (1877–1945) had done slightly earlier, but he also searched for what he called the "meta-hydrogen" of mass 3. Although his search was unsuccessful, he maintained that the corresponding nucleus consisting of three protons and two electrons ($^3_1H$ or $h_3e^+$ in his notation) was an important unit in the structure of the heavier atomic nuclei. The term "proton" as a synonym for the hydrogen nucleus or $H^+$ ion was only coined this year, and Harkins did not use it.[3] In lengthy and rather speculative papers in *Physical Review* and *Journal of the American Chemical Society*, Harkins argued that atomic nuclei consisted mainly of hydrogen and helium nuclei, meaning protons and alpha particles. However, he also considered other unit particles, including a mass-2 hydrogen nucleus ($^2_1H$) and two particles of mass 3, one being the $^3_1H$ particle and the other a nucleus consisting of three protons and one electron, that is, $^3_2He$. As to the former particle, the nucleus of meta-hydrogen, he wrote that "it would be the nucleus of an atom of a nuclear charge equal to one, and in this sense would be an isotope of hydrogen" [Harkins, 1920b, p. 83]. Harkins thought that $^3H$ might well be real but made up only a miniscule fraction of natural hydrogen. On the other hand, he speculated that it might be represented by the hypothetical element "nebulium" detected by unidentified spectral lines in the nebulae and sometimes ascribed an atomic weight of 2.7 [Hirsh, 1979].

Independently of Harkins, in his  of 1920, Rutherford reported to have found evidence of the $^3He$ nucleus in experiments with alpha particles colliding with gaseous molecules, such as $O_2$ and $N_2$. The

[3] For the origin of the name, see [Romer, 1997].

experimental data, he wrote, demonstrated the existence of "atoms of mass 3 carrying two positive charges ... [and with] physical and chemical properties very nearly identical with those of helium" [Rutherford, 1920, p. 394]. He suggested that the $^3$He isotope might be detected by means of the shift in its spectral lines, such as originally proposed by Bohr with respect to $^3$H.

Not only did Rutherford believe to have found $^3$He, he also found it "very likely that one electron can also bind two H nuclei and possibly also one H nucleus", which would entail "the possible existence of an atom of mass 2 carrying one charge ... an isotope of hydrogen" [Rutherford, 1920, p. 396]. Rutherford had considered the possibility of a mass-2 particle already in his early experiments with alpha particles colliding with nitrogen but without elaborating or commenting on their nature. On 10 January 1918, he entered in his notebook: "Suppose long-range scintillations in $N_2$ are due to atom charge + e and mass M = 2 called *x*" [Feather 1940, p. 152], and in his paper the following year, he cautiously suggested the existence of "atoms of mass 2" [Rutherford, 1919, p. 586]. While Rutherford thus anticipated the existence of $^3$He and $^2$H, he did not refer to the possibility of $^3$H.

The possible existence of two forms of triatomic hydrogen — one a molecule ($H_3$) and the other a heavy isotope ($^3$H) — led for a while to confusion, not least because Harkins, in some of his papers, insisted to designate his meta-hydrogen $H_3$. For example, in a report on inorganic chemistry to the Chemical Society, the British chemist Edward Baly (1871–1948) stated: "Now there seems no doubt that the helium isotope discovered by Rutherford is a different entity from $H_3$, which forms an integral part of Harkins's theory [and] was first discovered by Thomson. ... There thus exist two elements of mass 3, one an isotope of hydrogen and the other an isotope of helium" [Baly, 1921, p. 34]. The same confusion was expressed in an address to the British Association given by Edward Thorpe (1845–1925), an elderly chemist and historian of chemistry. According to Thorpe [1921, p. 292], "The hydrogen isotope $H_3$ [$^3$H], first detected by J. J. Thomson, of which the existence has been confirmed by Aston, would seem to be an integral part of atomic structure ... an isotope with an atomic weight of three and nuclear charge of one."

The chief advocates of the $H_3$ molecule, the American chemist Gerald Wendt (1891–1973) and his student Robert Landauer (1897–1966), considered the $^3H$ hypothesis in greater detail than previous authors. In this connection, they discussed the possibility of "iso-hydrogen", their name for the $^3H$ isotope. However, they argued that the isotope could not possibly be formed in the chemical reactions responsible for the formation of the $H_3$ molecule [Wendt and Landauer, 1922]. As we shall see, it would take a decade before the $^3H$ isotope returned to the scene of science.

## Digression: The Neutron as a Chemical Element

Both Rutherford and Harkins suggested around 1920 the existence of a bound proton-electron system, what they called a "neutron" and which differed in important respects from the later elementary particle of the same name. Rutherford [1920, p. 396] at first conceived the neutron — "an atom of mass 1 which has zero nucleus charge" — as if it were an atom of an unusual chemical element: "Its presence would probably be difficult to detect by the spectroscope, and it may be impossible to contain it in a sealed vessel". Given that the neutron had no external electron, it was a strange element indeed, and yet the German chemist Andreas von Antropoff (1878–1956), a professor at the University of Bonn, decided in 1926 to place it in his own version of the periodic table. As a name for this element of atomic number zero, he proposed "neutronium" [Fontana *et al.*, 2015, p. 444].

Antropoff's speculation was ignored, but after James Chadwick's (1891–1974) discovery of what we now call the neutron in 1932, it was revived by the ever-imaginative Harkins. Chadwick originally believed he found Rutherford's proton-electron particle, and it took more than a year before he realised that he had discovered a new elementary particle with a mass a little larger than the proton-plus-electron system. According to Harkins, Chadwick's neutron was the atom of an extraordinary form of matter, which he called "neuton" (he also considered the name "neutronium") and which he believed might be found in the interior of stars. In papers published in *Nature* and *Scientific Monthly*, he wrote enthusiastically about the "element zero, which is totally different from all the other elements ... [and] which possesses the unique characteristic of being at the same time an atomic nucleus and a complete atom" [Harkins, 1933]. Harkins' speculations of an element zero were not taken more seriously than those of Antropoff. When the astronomers Walter Baade (1893–1960) and Fritz Zwicky (1898–1974) in 1934 introduced the important idea of neutron stars, which were first observed in the form of pulsars in 1969, they were unaware of Harkins' suggestion.

## The Discovery of Deuterium

The hydrogen isotope of mass 2 was first considered in 1919, if only as a possibility. As mentioned above, in his paper of 1919, Rutherford referred briefly to the idea of a nucleus made up of two protons and one electron, an idea he had privately entertained the previous year. Slightly later and no less briefly, Harkins [1920b] suggested that atomic nuclei contained units of the same structure, or what in his terminology and nomenclature was "isohydrogen" (not to be confused with the iso-hydrogen that Wendt and Landauer considered) with the symbol $h_2e$. As yet another constituent of the nucleus, he proposed the neutral $h_2e_2$ unit, which in contemporary physics corresponds to a bound system of two neutrons or what is sometimes referred to as a "dineutron". The Rutherford-Harkins $h_2e$ particle corresponds to a deuteron.

Of more interest is a little-known contribution by two German scientists, the physicist Otto Stern (1888–1969) and the chemist Max Volmer (1885–1965). In a paper dated 7 January 1919, the two scientists analysed the deviations of the atomic weights of hydrogen and oxygen from whole numbers, which led them to consider the hypothesis of "an isotopic hydrogen nucleus of atomic weight 2 which possibly may be part of naturally occurring hydrogen" [Stern and Volmer, 1919, p. 226]. They calculated that in order to explain the atomic weights, the abundance of the heavy isotope should be approximately 0.8% of natural hydrogen. Moreover, they hypothesised that a mixed hydrogen molecule of molecular weight 3 might be responsible for what Thomson thought was $H_3$. The molecule they had in mind consisted of one ordinary hydrogen atom combined with an atom of the heavy isotope. Stern and Volmer reported elaborate diffusion experiments with the purpose of detecting the heavy hydrogen isotope but without finding any trace of either this isotope or isotopes of oxygen. "We have thus proved," they concluded, "that ... hydrogen and oxygen are not mixtures of isotopes" [Stern and Volmer, 1919, p. 238]. In reality, their experiments merely showed that for every million light hydrogen atoms, there were less than 10 heavy atoms.

Other researchers in the 1920s agreed with the negative conclusion of Stern and Volmer. Thus, in his authoritative monograph *Isotopes*, Aston [1924, p. 72] stated confidently that "it is safe to conclude that hydrogen is a simple element" and that the same was the case with oxygen. Aston's "simple element" was one composed of just a single isotope. Only in 1929 did the American chemist William Giauque (1895–1982) and his collaborator Herrick Johnston (1898–1965) discover oxygen isotopes with mass numbers 17 and 18 in the atmosphere. It took another two years before the hydrogen mass-2 isotope was taken seriously. Since the discovery of $^2H$ or deuterium is well described in the literature [Brickwedde, 1982; Coffey, 2008, pp. 208–221; Clark and Reader, 2012], I only recapitulate it in outline.

The Giauque-Johnston discovery of heavy oxygen isotopes indirectly paved the way for the discovery of deuterium [Shindell, 2019, p. 75]. In May 1931, the physicist Raymond Birge (1887–1980) and the astrophysicist Donald Menzel (1901–1976) referred to the possibility of a mass-2 hydrogen isotope, which they thought might explain why the atomic weight of the element came out with the same value on the physical and the chemical mass scale. Whereas the physicists adopted the standard $^{16}O = 16$, the International

Commission of Atomic Weights assigned the atomic weight 16 to natural oxygen [Holden, 2004]. The recognition that naturally occurring oxygen is a mixture of $^{16}O$,$^{17}O$ and $^{18}O$ thus implied a small difference between the physical and chemical atomic weight scales. The idea of Birge and Menzel was essentially the same as the one examined more than a decade earlier by Stern and Volmer, but Birge and Menzel were apparently unaware of the German paper. What matters is that the weight numbers must be slightly different on the two scales and yet precision measurements gave the same value of hydrogen's atomic weight whatever scale used. By 1931, the best chemical value was 1.00777 ± 0.00002 atomic mass units, and the physical value was 1.00778 ± 0.00015. It was in this context that Birge and Menzel [1931, p. 1670] pointed out that the coincidence might be explained "by speculating the existence of an isotope of hydrogen of mass 2, with a relative abundance $H^1/H^2 = 4500$".

Harold Clayton Urey (1893–1976), an associate professor of chemistry at Columbia University since 1929, was an expert in quantum theory and physical chemistry [Shindell, 2019]. As evidenced by a massive textbook of 1930 coauthored with Arthur Ruark (1899–1979) and titled *Atoms, Molecules and Quanta*, Urey did not presume hydrogen to be composed of more than one isotope at the time. The two authors briefly referred to the old paper by Stern and Volmer, confidently accepting their claim that hydrogen was not a mixture of two or more isotopes. However, the Birge-Menzel paper, which appeared the year after, caused Urey to change his mind. Immediately after having read the paper, he decided to transform the speculation into reality by looking systematically for the $^2H$ isotope. For theoretical reasons, he suspected that there existed nuclei with two and three protons bound together by one and two electrons, respectively, or, in modern language, the nuclei of deuterium and tritium; he also suspected a helium isotope of mass 5 (Figure 1). Given that the argument of Birge and Menzel turned out to be invalid, as it rested on a slightly wrong value of hydrogen's atomic weight, it is ironic that their incorrect paper served as a direct inspiration for Urey's discovery

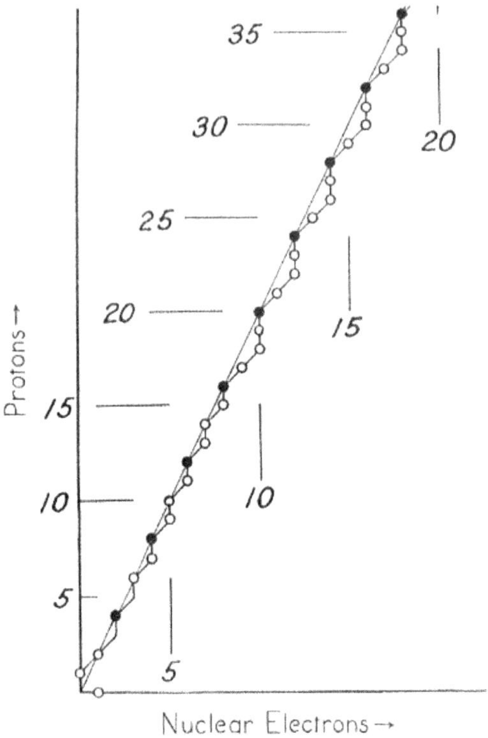


Figure 1: Urey's 1931 chart of protons versus nuclear electrons, with filled circles indicating known isotopes and open ones indicating those that would be expected to be found according to the trend shown in the chart. Source: [Urey *et al.*, 1932, p. 14]. Reprinted with permission from the American Physical Society.

of deuterium. As Urey [1966, p. 354] (see also [Meija, 2022]) stated in a postscript to his Nobel Lecture, "it is probable that ... the discovery of deuterium would have been delayed for some time" without the error.

To detect the as-yet hypothetical $^{2}H$ isotope, Urey teamed up with his Columbia colleague George Murphy (1904–1969). Their plan was to look for the characteristic spectral shift in samples of liquid hydrogen prepared by means of fractional distillation by Ferdinand Brickwedde (1903–1989), a physicist at the National Bureau of Standards in Washington DC. In a careful investigation of the spectral lines, Urey and Murphy found shifts in wavelengths, which they attributed to a $^{2}H$ component. Since the measured shifts agreed with those calculated, they were confident that they had now discovered the mass-2 isotope.[4]

The discovery process concerning the new $^{2}H$ discovery was planned, systematic, and with no surprises. In the last days of 1931, Urey reported the results in a ten-minute talk at a meeting of the American Physical Society, and the detailed paper of Urey and his collaborators appeared in *Physical Review* on 1 April 1932. According to the three authors, the relative abundance of the new isotope was about 1:4000, of the same order as the presently known ratio 1:6700. Urey, Murphy and Brickwedde [1932, p. 15] not only looked for $^{2}H$ but also for $^{3}H$. However, "no evidence for $H^3$ has yet been found, but further concentration ... may yet show that this nuclear species exists". The three authors did not discuss the constitution of the $^{2}H$ nucleus but implicitly assumed that it was composed by two protons and one electron. After Chadwick's discovery of the neutron later in 1932, the accepted constitution of the deuteron changed from $(2p^+, e^-)$ to $(p^+, n)$.

Whereas the discovery process went smoothly, the naming of the heavy isotope and its nucleus did not. On the contrary, it involved a protracted debate [Stuewer, 1986]. A variety of names were proposed and discussed, among them "diplogen" and "diplon", which Rutherford preferred, while others argued that the $^{2}H$ nucleus should be called a "deuton", a name frequently used in the period. Still, another suggestion was "dygen"/"dyon", as argued by some American scientists. Urey, on the other hand, was in favour of deuterium/deuteron, and for the other two hydrogen isotopes, he proposed "protium" ($^{1}H$) and "tritium" ($^{3}H$). Although protium is rarely used, Urey's names eventually won official recognition [IUPAC, 1971].

The discovery of deuterium was quickly confirmed by other physicists and chemists, and within a year, more than a hundred

---

[4]The strongest of the hydrogen lines has a wavelength of 6,563 Å (angstrom). For this line, Urey and Murphy observed a spectral shift of 1.79 Å and calculated 1.787 Å on the assumption that it was due to the presence of a mass-2 isotope.

research papers were published on the new isotope and its chemical compounds. Urey's former professor, the prominent Berkeley chemist Gilbert Newton Lewis (1875–1946), was among those who jumped on the bandwagon, quickly writing a series of papers on deuterium and heavy water. According to Urey's biographer, Lewis's sudden activity in deuterium studies was resented by Urey, who felt the competition from Lewis to be unfair [Shindell, 2019, p. 77; see also below].

The importance of the discovery was recognised by the Royal Swedish Academy of Sciences, which, in 1934, awarded Urey the full Nobel Prize in Chemistry "for his discovery of heavy hydrogen". The spectroscopic method that led to the discovery had its origin in Bohr's old theory, although by 1931, this theory had been replaced by the new atomic theory based on quantum mechanics. Actually, the latter theory gives the very same expression for the isotope shift as the older one. In 1923, Urey had studied for a year at Bohr's institute in Copenhagen, but he most likely was unaware of Bohr's early and half-forgotten prediction of the isotope effect. In his Nobel lecture, Urey [1966, p. 342], however, mentioned that "Bohr's theory, given some twenty years ago, permits the calculation of the Balmer spectrum of the heavier isotopes of hydrogen by the well-known theoretical formula for the Rydberg constant" (Figure 2).

Although Urey's work was generally hailed as one of the great discoveries of the 20$^{th}$ century, not all experts were happy with designating deuterium as an isotope of hydrogen. In a discussion meeting of the Royal Society, none other than Frederick Soddy (1877–1956), who on 4 December 1913 had introduced the notion of isotopy and in 1921 was rewarded with a Nobel Prize for his work, objected that chemical inseparability was a key criterion of isotopy. Without accepting the modern definition of isotopy in terms of nuclear structure, he insisted that deuterium was not an isotope of hydrogen but merely a new species of the element. Speaking of "the questionable extension of the now well-understood term 'isotope' to heavy hydrogen", Soddy stated that "hydrogen isotope" was simply a misnomer [Rutherford *et al.*, 1934, p. 11]. As he saw it, since the

Figure 2: Bohr and Urey in conversation at a meeting in Zurich in 1957. Credit: Niels Bohr Archive Photo Collection.

light and heavy hydrogen could be separated chemically, they were not isotopes of the same element.

## The Roads to Tritium

Although briefly discussed by Harkins and a few other scientists around 1920, the hypothetical $^3H$ isotope was not taken very seriously at first and seems to have been ignored until the early 1930s, when interest was reawakened by the discovery of deuterium. At about 1930, the American physicist Fred Allison (1882–1974) developed a magneto-optical method of chemical analysis by means of which he and his coworkers claimed to have made a number of remarkable discoveries, which included no less than 16 lead isotopes and, most notably, the discovery of elements 85 and 87, which he named "alabamium" and "virginium", respectively

[Scerri, 2013, pp. 149–152]. Allison [1933, p. 75] also claimed to have found evidence of a heavy hydrogen isotope in 1931 before the work of Urey. However, Allison's method and his use of it were controversial, and by the mid-1930s, his work was generally disbelieved and eventually debunked as pseudoscience [Fontani *et al.*, 2015, pp. 327–331]. Indeed, when the American physical chemist Irving Langmuir (1881–1957), a Nobel Prize laureate of 1932, introduced the concept of "pathological science" as another name for pseudoscience, Allison's work featured as a prime example [Langmuir, 1989].

Before Allison was debunked, the Berkeley chemist Wendell Latimer (1893–1955) made magneto-optical experiments with water solutions containing about 3% deuterium, reporting that he had found evidence of $^{3}$H [Latimer and Young, 1933]. However, the claim was contradicted by careful spectrographic experiments with 67% concentrated heavy water made by G. N. Lewis and his assistant Frank Spedding (1902–1984). Failing to find any trace of the expected spectral shift within the sensitivity of their apparatus, the two chemists concluded that the $^{3}$H isotope "appears not to be present in ordinary hydrogen to the extent of 1 part in 6,000,000" [Lewis and Spedding, 1933, p. 964]. Nor did painstaking studies with the mass spectrometer reveal any sign of the superheavy hydrogen isotope.

If $^{3}$H was absent from water and also from the solar spectrum (where Lewis and Spedding had also looked for it), perhaps it might turn up in nuclear reactions? British and American physicists thought so and in the spring of 1934, the plot thickened with several claims of having detected the elusive particle. By that time, the atomic nucleus no longer consisted of protons and electrons but of protons and neutrons. At the Cavendish Laboratory, Rutherford and his assistants, the Australian physicist Mark Oliphant (1901–2000) and the Austrian physical chemist Paul Harteck (1902–1985), bombarded deuterium compounds with deuterons accelerated in the laboratory's new high-voltage generator. The experiments were principally due to Oliphant (Figure 3) and Harteck, while Rutherford's

Figure 3: Mark Oliphant, co-discoverer of tritium, at the cafeteria of Bohr's institute during a conference in 1936. The person to his right is Czech physicist George Placzek. Credit: Niels Bohr Archive Photo Collection.

contribution was largely limited to discussion and interpretation of the results. The three researchers suggested as one possibility the reaction

$$^{2}_{1}\mathrm{H} + {}^{2}_{1}\mathrm{H} \rightarrow {}^{4}_{2}\mathrm{He}^{*} \rightarrow {}^{1}_{1}\mathrm{H} + {}^{3}_{1}\mathrm{H}$$

with the release of 4.0 MeV of energy. The second possibility was that the unstable $^{4}\mathrm{He}^{*}$ nucleus decayed into a He-3 nucleus and a neutron according to

$$^{2}_{1}\mathrm{H} + {}^{2}_{1}\mathrm{H} \rightarrow {}^{4}_{2}\mathrm{He}^{*} \rightarrow {}^{3}_{2}\mathrm{He} + {}^{1}_{0}n$$

In this case, the surplus energy would be 3.2 MeV. Not only did Rutherford and his two coworkers believe they had detected the new isotopes $^{3}\mathrm{H}$ and $^{3}\mathrm{He}$, but they also thought they had found a stable Be-8 isotope in some of the nuclear reactions. (This isotope is, in fact, exceedingly unstable, with a lifetime of only $10^{-16}$ s.)

The Cavendish group published a preliminary report in *Nature* on 17 March 1934, in which they cautiously stated that the two mass-3 isotopes "appear to be stable for the short time required for their detection, [but] the question of their permanence requires further consideration" [Oliphant *et al.*, 1934a]. Rutherford judged the potential discoveries of the two isotopes in somewhat different terms. As he confided in a letter to Ernest Lawrence (1901–1958) on 13 March 1934, "I personally believe that there can be little doubt of the reaction in which the hydrogen isotope of mass 3 is produced, ... [while] the evidence for the helium isotope of mass 3 is ... at present somewhat uncertain" [Dahl, 2002, p. 155]. In their detailed paper in the *Proceedings of the Royal Society* dated a month later, Rutherford and his collaborators similarly expressed confidence in the discovery of $^3$H, whereas their evaluation of the mass-3 helium isotope was more circumspect: "While we have not yet detected the $_2$He$^3$ particles which we believe to be present, we have not yet obtained any evidence that they do not exist" [Oliphant *et al.*, 1934b, p. 702].

American physicists were competing with and closely following the work done at the Cavendish Laboratory. They might have been beaten in the race to find the $^3$H isotope, but what about its presence in nature? In April 1934, a group of three Princeton physicists analysed a sample of nearly pure deuterium with an advanced mass spectrometer and reported evidence of a mass-5 component, which they interpreted as due to a binary molecule consisting of one $^2$H atom (D) and one $^3$H atom (T). Without using the name "tritium", they designated it the symbol "T". The estimated abundance was exceedingly low, namely about one $^3$H atom per one billion of $^1$H atoms. Nonetheless, the result "confirms rather satisfactorily the existence of a third isotope of hydrogen from natural sources" [Lozier *et al.*, 1934].

Shortly later, another team of American physicists reported evidence for $^3$H atoms, this time based on the study of nuclear reactions roughly of the same kind as those made at the Cavendish Laboratory. Merle Tuve (1901–1982) and his coworkers at the Carnegie Institution in Washington DC used high-speed deuterons to bombard a highly concentrated deuterium target sample, and by measuring the range of the resulting particles, they found evidence for $^3$H atoms, whereas Rutherford's group concluded that the $^3$H nuclei were the result of a nuclear transformation. Tuve's group suggested that they were stable constituents present in the $^2$H target and consequently, that their work provided "satisfactory proof of the existence of $H^3$ atoms in deuterium samples" [Tuve *et al.*, 1934, p. 841]. Unfortunately, nobody else could reproduce the results communicated by Tuve and his group.

The situation in the fall of 1934 was summarised in a popular paper with the title "Protium-deuterium-tritium: The hydrogen trio" written by Hugh Taylor (1890–1974), a chemistry professor at Princeton University. According to Taylor [1934, p. 367], there was no doubt that tritium — "the youngest and rarest of all the isotopes" — was part of the natural world. But even so, tritium compounds would never become commodities of any practical use: "Pure tritium water, $T_2O$, would cost at least $10,000,000 a gram or [tritium] water roughly twenty times the cost of radium".

In a paper published in *Nature* on 21 August 1937, which turned out to be his last one, Rutherford carefully appraised the history and current status of $^3$H and $^3$He, referring to $^3$H as "triterium" rather than tritium. He maintained the Cavendish priority to the discoveries of the two isotopes but now concluded that they rapidly disappeared in experiments and that none of them existed naturally. Yet he did not suspect any of the isotopes to be radioactive. Upon the instigation of Rutherford, Aston used his mass spectrograph to look for the $^3$H isotope in a sample of pure heavy water supplied by the Norwegian company Norsk Hydro. However, contrary to what Rutherford expected, Aston's careful analysis showed no sign of the mass-3 isotope. Rutherford [1937, p. 305] commented: "It does not seem feasible at the moment to obtain sufficient quantities of these two interesting isotopes to study properties by ordinary physical chemical methods." This was not the end of the discovery history, for it can be argued that tritium was not discovered in 1934 but only five years later.

In 1939, the young Berkeley physicist Luis Alvarez (1911–1988), who 29 years later would be awarded the Nobel Prize for his contributions to elementary particle physics, and his graduate student Robert Cornog (1912–1998) used the Berkeley cyclotron to produce a steady beam of 24-Mev $^3He$ ions. They demonstrated that traces of the rare isotope were present in ordinary helium and estimated an abundance ratio relative to atmospheric helium of about $10^{-7}$. Alvarez and Cornog [1939] (see also [Cornog, 1987]) further showed that the gas mixture produced in the deuteron-deuterium process was radioactive, which they ascribed to a $^3H$ component decaying into $^3He$. On the other hand, they found the $^3He$ nucleus to be stable. Alvarez and Cornog also investigated the half-life of tritium in their 1939 experiments, but without coming up with a definite result. The first determination of tritium's half-life followed in 1940, yielding 31 ± 8 years, which is in fair agreement with the modern value of 12.3 years. Deuterium, as well as tritium, proved immensely important, but the later consequences of the discoveries related to thermonuclear weapons are outside the scope of the present paper [Rhodes, 1995].

## Discussion

Whereas the discoveries of new chemical elements are officially recognised either by IUPAC or informally by the chemical community, this is not generally the case with the numerous isotopes belonging to the elements. Just as astronomers do not celebrate the discovery of yet another comet, chemists and physicists consider the discovery of yet another isotope to be relatively unimportant. There are too many of them to make headlines in the history of science. While the discovery of a new element has in several cases been recognised with a Nobel Prize, witness radium, polonium, plutonium, and the inert gases, only a single isotope discovery has been similarly rewarded, namely, Urey's discovery of deuterium. The discovery was fairly unproblematic and has never been seriously questioned nor confronted with rival discovery claims. Urey and his collaborators Murphy and Brickwedde did not refer to their work on deuterium as a "discovery" in any of their papers of 1932, but the Nobel Committee had little doubt that it qualified as one.

Although Urey's paternity to deuterium was broadly accepted, his full Nobel Prize for the discovery was far from self-evident. There was another serious candidate for the prize, namely G. N. Lewis, whose important work on heavy water and deuterium chemistry was highly considered by the Nobel Committee. In a report to the committee of 18 May 1934, Theodor Svedberg (1884–1971), a reputed Swedish physical chemist and himself a Nobel laureate of 1926, stated that "the main merit for the discovery is without doubt ascribed to Urey" [Coffey, 2008, p. 217]. Nonetheless, Svedberg ended up recommending the prize be shared by Urey and Lewis. Only after further deliberations did he change his mind, now arguing that Lewis's work was, after all, derivative to that of Urey. To the disappointment of Lewis, when the prize was announced on 15 November 1934, it went to Urey alone [Coffey, 2008, p. 221; Shindell, 2019, p. 78]. It could have been divided between him and his two collaborators, Murphy and Brickwedde, but this was not contemplated in Stockholm. Apparently, Urey thought they deserved credit for their contributions to the discovery, for he generously gave half of his Nobel Prize money to be divided between them [Coffey, 2008, p. 2019; Shindell, 2019, p. 81].

The *idea* of a heavy hydrogen isotope of mass 2 was not new, as it was discussed on a few occasions around 1920, after which it largely disappeared. Rutherford and Harkins independently hypothesised the existence of a mass-2 nuclear particle, but their suggestions were speculative and unconnected to potential experiments. The only pre-discovery experiments aimed at the $^{2}H$ isotope were made in Berlin by Stern and Volmer, who concluded that the isotope did not exist. In agreement with what has been argued in the first part of this chapter, none of these early cases can reasonably be characterised as predictions of the discovery made in 1932. They were merely loose speculations.

As far as tritium is concerned, Bohr suggested in 1913 that the triatomic hydrogen molecule proposed by Thomson in his positive-ray experiments might be a $^{3}H$ isotope. In an interview on 1 November 1962, Bohr recalled his role in the prehistory of tritium: "I just took up the question of whether in hydrogen one could have what you now call tritium. And then I saw that it was a way to show this by its diffusion in palladium. Hydrogen and tritium will behave similarly but the masses are so different that they will be separated out" [Kragh, 2012b, p. 178]. Although the suggestion was scarcely noticed at the time, it was made in public and counts as a proper prediction. On the other hand, Bohr's slightly later and more important idea of detecting $^{3}H$ by spectroscopic means was not known to the scientific community and is, therefore, irrelevant in a social discovery context. And yet it is of interest retrospectively, given that this "private discovery claim" might, under other circumstances, have been communicated to the scientists and thus constituted a proper discovery claim. Insofar that Harkins in 1920 presented his idea of "meta-hydrogen" as a mass-3 isotope and actually looked for it, he may be considered as having anticipated tritium.

The case of tritium is significantly different from the one of deuterium, not only because of the scarcity of this unstable and more exotic isotope but also because there are two possible candidates for having discovered tritium. According to *Encyclopaedia Britannica* and several other sources, the $^{3}H$ isotope was discovered by Rutherford, Oliphant and Harteck in 1934, but there are also sources that credit Alvarez and Cornog with the discovery. The 1939 paper by Alvarez and Cornog did not present their work as a discovery of a new isotope, and yet, in his recollections many years later, Cornog [1987, p. 26] stated that he and Alvarez "discovered hydrogen and helium of mass 3".

Was tritium discovered twice? There are indeed cases in which elements or something else can reasonably be said to have been discovered twice or even thrice (such as the discovery of chlorine or aluminium),[5] but in my view, tritium does not belong to this category. As shown by the historical record, the Cavendish group demonstrated the existence of the $^{3}H$ nucleus in their experiments of 1934, and what is no less important, the claim was generally accepted by the relevant scientific communities. It is worth noting that, strictly speaking, the Cavendish group did not discover the atomic species tritium but only its nucleus, sometimes known as a triton. In this respect, there is a

[5] See the contributions by Pere Grapi and Sarah Hijmans in this volume.

parallel to the much later artificially produced "superheavy elements" of atomic number $Z > 103$. These have, in some cases, been recognised as real and hence being discovered on the basis of the detection of only a few atomic nuclei [Kragh, 2018].[6] There is also an interesting analogy to the discovery of technetium ($Z = 43$), which, in 1937, was synthesised by Emilio Segrè (1905–1989) and Carlo Perrier (1886–1948) as the first-ever element produced in the laboratory.[7] It is often forgotten that tritium was similarly produced three years earlier. The analogy goes further: as tritium is not totally absent in nature, so is the case with technetium.

As far as the work of Alvarez and Cornog is concerned, the two physicists undoubtedly made a discovery, but not of a new hydrogen isotope. They discovered that tritium is radioactive and also that the $^3$He isotope is part of natural helium, but with regard to tritium, they merely confirmed its existence. It is hard to avoid the conclusion that priority to the discovery of tritium belongs to Rutherford and his group at the Cavendish Laboratory. Nonetheless, it was only with the Alvarez-Cornog work that tritium became fully recognised as an important object in nuclear physics. The discovery histories of the heavy hydrogen isotopes illustrate the problematic relationship between the related yet different concepts of anticipation, prediction and discovery. In the cases of both deuterium and tritium, there were earlier suggestions and theoretically based expectations of the possible existence of the isotopes. The first group may qualify as anticipations and the second as predictions, but whatever their names, they were not discoveries. In my view, they missed the crucial element of experimental detection widely accepted by the scientific community.

## References

Allison, F. (1933). The magneto-optic method of analysis. *Journal of Chemical Education*, 10(2), pp. 71–78.

[6] See also Ann Robinson's contribution to this volume.

[7] See also Brigitte Van Tiggelen's contribution to this volume.

Alvarez, L. W. and Cornog, R. (1939). Helium and hydrogen of mass 3. *Physical Review*, **56**, p. 613.

Aston, F. (1924). *Isotopes* (Edward Arnold & Co., London).

Baly, E. C. (1921). Inorganic chemistry. *Ann Progr Rep Chem Soc*, **17**, pp. 27–56.

Birge, R. T. and Menzel, D. H. (1931). The relative abundance of the oxygen isotopes, and the basis of the atomic weight system. *Physical Review*, **37**, pp. 1669–1671.

Brickwedde, F. G. (1982). Harold Urey and the discovery of deuterium, *Physics Today*, **35**(9), pp. 34–39.

Clark, C. W. and Reader, J. (2012). Light, atoms and nuclei: The optical discovery of deuterium. *Optics & Photonics News*, **23**(5), pp. 36–41.

Coffey, P. (2008). *Cathedrals of Science: The Personalities and Rivalries that Made Modern Chemistry* (Oxford University Press, Oxford).

Connolly, N. G. and Damhus, T. (2005). *Nomenclature of Inorganic Chemistry. IUPAC Recommendations 2005* (RCS Publishing, Cambridge).

Cornog, R. (1987). Discovery of hydrogen helium three. In *Discovering Alvarez*, ed. Trower, W.P. (University of Chicago Press, Chicago), pp. 26–28.

Dahl, P. F. (2002). *From Nuclear Transmutation to Nuclear Fission 1932–1939* (Institute of Physics Publishing, Bristol).

Eve, A. S. (1939). *Rutherford. Being the Life and Letters of the Rt Hon. Lord Rutherford, O.M.* (Cambridge University Press, Cambridge).

Feather, N. (1940). *Lord Rutherford* (Blackie & Sons, London).

Fontani, M., Costa, M. and Orna, M. V. (2015). *The Lost Elements: The Periodic Table's Shadow Side* (Oxford University Press, Oxford).

Harkins, W. D. (1920a). The separation of the element chlorine into chlorine and meta-chlorine, *Science*, **51**, pp. 289–291.

Harkins, W. D. (1920b). The nuclei of atoms and the new periodic system. *Physical Review*, **15**, pp. 73–94.

Harkins, W. D. (1933). The new kind of matter: Element zero or neuton. *Sci Mon*, **36**, pp. 546–549.

Hirsh, R. F. (1979). The riddle of the gaseous nebulae. *Isis*, 70, pp. 197–212.

Holden, N. E. (2004). Atomic weights and the International Committee — A historical review. *Chemistry International*, **26**, pp. 4–7.

IUPAC (1971). *Nomenclature of Inorganic Chemistry*, 2nd ed. (Butterworths, London).

Kragh, H. (2012a). To be or not to be: The early history of $H_3$ and $H_3^+$, *Philosophical Transactions of the Royal Society A*, **370**, pp. 5225–5235.

Kragh, H. (2012b). The isotope effect: Prediction, discussion, and discovery. *Studies in History and Philosophy of Modern Physics*, **43**, pp. 176–183.

Kragh, H. (2018). *From Transuranic to Superheavy Elements: A Story of Dispute and Creation* (Springer, Cham, Switzerland).

Kragh, H. (2019). Controversial elements: Priority disputes and the discovery of chemical elements. *Substantia*, **3**(2, Suppl. 5), pp. 79–90.

Langmuir, I. (1989). Pathological science. *Physics Today*, **42**, pp. 36–48.

Latimer, W. M. and Young, H. A. (1933). The isotopes of hydrogen by the magneto-optic method. The existence of $H^3$. *Physical Review*, **44**, p. 690.

Lewis, G. N. and Spedding, F. H. (1933). A spectrographic search for $H^3$ in concentrated $H^2$. *Physical Review*, **43**, pp. 964–966.

Lozier, W. W., Smith, P. T. and Bleakney, W. (1934). $H^3$ in heavy hydrogen. *Physical Review*, **45**, p. 655.

Meija, J. (2022). The unit that shall not be named. *Nature Physics*, **18**, p. 602.

Oliphant, M., Harteck, P. and Rutherford, E. (1934a). Transmutation effects observed with heavy hydrogen, *Nature*, **133**, p. 413.

Oliphant, M., Harteck, P. and Rutherford, E. (1934b). Transmutation effects observed with heavy hydrogen. *Proceedings of the Royal Society A*, **144**, pp. 692–703.

Rancke-Madsen, E. (1975). The discovery of an element. *Centaurus*, **19**, pp. 299–313.

Rhodes, R. (1995). *Dark Sun: The Making of the Hydrogen Bomb* (Simon & Schuster, New York).

Romer, A. (1997). Proton or prouton? Rutherford and the depths of the atom. *American Journal of Physics*, **65**(8), pp. 707–716.

Rutherford, E. (1919). Collision of α particles with light atoms, IV: An anomalous effect in nitrogen. *Philosophical Magazine*, **37**, pp. 581–587.

Rutherford, E. (1920). Nuclear constitution of atoms. *Proceedings of the Royal Society A*, **97**, pp. 374–400.

Rutherford, E. (1937). The search for the isotopes of hydrogen and helium of mass 3. *Nature*, **140**, pp. 303–305.

Rutherford, E., *et al.* (1934). Discussion on heavy hydrogen. *Proceedings of the Royal Society A*, **144**, pp. 1–28.
Scerri, E. (2013). *A Tale of 7 Elements* (Oxford University Press, Oxford).
Shindell, M. (2019). *The Life and Science of Harold C. Urey* (University of Chicago Press, Chicago).
Stern, O. and Volmer, M. (1919). Sind die Abweichungen der Atomgewichte von der Ganzzahligkeit durch Isotopie erklärbar? *Annalen der Physik*, **59**, pp. 225–238.
Stuewer, R. H. (1986). The naming of the deuteron. *American Journal of Physics*, **54**, pp. 206–218.
Taylor, H. S. (1934). Protium-deuterium-tritium: The hydrogen trio. *Sci Mon*, **39**, pp. 364–372.
Thorpe, T. E. (1921). The constitution of matter. *Sci Mon*, **13**, pp. 289–300.
Tuve, M. A., Hafstad, L. R. and Dahl, O. (1934). A stable isotope of mass 3. *Physical Review*, **45**, pp. 840–841.
Urey, H. C. (1966). Some thermodynamic properties of hydrogen and deuterium. In *Nobel Lectures Chemistry 1922–1941* (Elsevier, Amsterdam), pp. 339–354.
Urey, H. C., Brickwedde, F. C. and Murphy, G. M. (1932). A hydrogen isotope of mass 2 and its Concentration. *Physical Review*, **40**, pp. 1–15.
Wendt, G. and Landauer, R. (1922). Triatomic hydrogen. *Journal of the American Chemical Society*, **44**, pp. 510–521.

## Selected Bibliography

Brickwedde, F. G. (1982). Harold Urey and the discovery of deuterium. *Physics Today*, **35**(9), pp. 34–39.
Coffey, P. (2008). *Cathedrals of Science: The Personalities and Rivalries that Made Modern Chemistry* (Oxford University Press, Oxford).
Kragh, H. (2012). The isotope effect: Prediction, discussion, and discovery. *Studies in History and Philosophy of Modern Physics*, **43**, pp. 176–183.
Schimmelmann, A. and Sauer, P. (2018). Hydrogen isotopes. In *Encyclopedia of Geochemistry*, ed. White, W. M. (Springer, Cham, Switzerland).
Shindell, M. (2019). *The Life and Science of Harold C. Urey* (University of Chicago Press, Chicago).

# Towards a New Model for Discovery History

https://doi.org/10.1142/9789819813858_0010

# IN DEFENCE OF A JOURNALISTIC MODEL OF DESCRIBING DISCOVERY: THE CASES OF ARGON, HELIUM AND NEON

Carmen J. Giunta

*Department of Chemistry, Le Moyne College, Syracuse, NY, USA.*
*Giunta@lemoyne.edu*

The nature of scientific discovery is a topic much investigated by historians, philosophers, and sociologists of science, and it is a topic of interest to scientists and science educators as well. The present volume focuses on a subset of examples within a single discipline, the discoveries of chemical elements. This set of roughly 100 examples constitutes a rich database from which descriptive models of discovery can be constructed or normative sets of criteria tested.

In this chapter, I describe several such models, from the justly criticised individualist, or "heroic" model, to approaches that question the possibility of demarcating discoveries. I examine them in light of the discoveries of three elements — argon, helium and neon — during the final decade of the $19^{th}$ century. I selected these elements because they include examples whose treatment by other authors [Arabatzis and Gavroglu, 2016; Kragh, 2009; 2019] I wish to review, and because they display characteristics (such as predictable or unpredictable) relevant to the models I discuss. The common journalistic questions of "who", "what", "when", and "where" are often associated with the individualist model of discovery; however, I suggest that these questions are not responsible for the model's shortcomings. Furthermore, I contend that these questions, along with their colleagues, "how" and "why", form a suitable basis for describing discoveries of elements, so long as simple answers are not assumed or demanded. I call a framework for describing scientific discoveries in terms of answers to these questions a "journalistic model".

## Models of Element Discovery

In 1975, the Danish chemist Edmond Rancke-Madsen (1910–1987) proposed a set of criteria to answer the question, "What qualifies a person to be regarded as the discoverer of an element?" He formulated two criteria to identify the "effective discoverer" of an element, the first of which deals with experimental or observational aspects and the second with announcing results. The discoverer of an element

> ... has observed the existence of a new substance which is different from earlier described substances, and this new substance is recognized by him [*sic*] or later by scientists as being elemental. The element in question may exist in chemical combination or as a free element, without any demand of purity being made of this particular substance.
>
> [Rancke-Madsen, 1975, p. 300]

The discoverer must also announce results to an audience that extends beyond her or his immediate associates, and that announcement must be "noticed". The announcement completes the process of discovery, and the date of announcement is considered the date of discovery.

Rancke-Madsen also proposed that the chronology of the discovery of an element can often be divided into three phases: a preliminary phase, the effective discovery, and subsequent developments. A preliminary phase may include a time when a new element is suspected or predicted; it might also be a time when the element was "ineffectively discovered" (my term, not his) in a way that was not communicated or not noticed. The phase of effective discovery is that time during which the novel characteristics of the substance in question are "recognised in a convincing manner and published" [Rancke-Madsen, 1975, p. 301]. Subsequent developments (which, it seems to me, belong not to the discovery proper but to the further characterisation of the element) may include preparation of the element, description of its physical and chemical properties, and detection of allotropes or isotopes. Rancke-Madsen did not list the recognition of elementary character in this phase, but logically, it belongs here if it was not done before the end of the previous phase.

In a paper concerned with priority disputes in the discovery of chemical elements, the Danish historian of science Helge Kragh [2019] reviewed several formulations of scientific discovery in general and the discovery of elements in particular before articulating his understanding of the discovery of chemical elements:

> The discovery of a new element is thus to demonstrate convincingly the elemental nature of some substance, which is a conceptual discovery, and also to find this substance in nature — or perhaps to synthesize it in the laboratory. The latter is an empirical discovery.
>
> [Kragh, 2019, p. 80]

Kragh nodded to Rancke-Madsen's notion of "effective discovery" by adding, "a discovery cannot be a private matter". He acknowledged that this description of the discovery of an element focuses on intellectual history while neglecting social history. The two are complementary, he noted, and neither approach alone tells the entire story.

Although working in an intellectual history mode, Kragh described and criticised another intellectual-historical model of scientific discovery, one which he called the individualist or "heroic" model [Kragh, 2019] or the "mentalist" view [Kragh, 2009]. Theodore Arabatzis and Kostas Gavroglu called it the "received view of scientific discovery" [Arabatzis and Gavroglu, 2016]. They characterised it by the questions of "who", "what", "when", and "where"? It regards discoveries as discrete events occurring at identifiable locations and times, attributable to individual investigators, and involving identifiable discovered objects or entities. The individualist model is common among journalists and scientists, Kragh [2019] and Arabatzis and Gavroglu [2016] noted.

Historians have criticised this model for decades. As was pointed out by Arabatzis and Gavroglu [2016], Thomas Kuhn (1922–1996) exposed some of its weaknesses as long ago as 1962 in a short paper published in *Science*. There, Kuhn distinguished between predictable and unpredictable discoveries. The former are discoveries of entities expected on the basis of currently accepted knowledge. Kuhn's examples included neutrinos and elements that matched empty spaces in the periodic table; more recent examples would include gravitational waves and superheavy elements. In unpredictable discoveries, "there are no benchmarks to inform either the scientist or the historian when the job of discovery has been done" [Kuhn, 1962, p. 761]. A discovery that is truly novel or even apparently contradictory to accepted knowledge (and therefore *a fortiori* unpredictable) involves "recognising both *that* something is and *what* it is" [Kuhn, 1962, p. 762; emphasis in original]. Such a discovery begins with noticing an anomaly and then "making the anomaly behave" [Kuhn, 1962, p. 763]. The discovery is not complete until it is assimilated into the accepted theoretical framework of the science, thereby altering the previously accepted framework.

Arabatzis and Gavroglu [2016] examined the discovery of argon in light of Kuhn's ideas. They noted that the standard questions of the individualist model did not have unique answers. Their principal assertion was that before argon could be recognised as an element, the unexpected properties of the new substance had to be assimilated into the concepts of the science of the time. This assimilation included adding the possibility of inertness to the concept of a chemical element and finding a place for argon in the periodic table, which was eventually altered by the insertion of a new group.

Before leaving the individualist model of discovery, I note that even its critics found that it was sometimes adequate. Kuhn [1962, p. 761] asserted that in the case of predictable discoveries, "only a paucity of data can prevent the historian from ascribing them to a particular time and place". Kragh also noted that the questions that the model presupposes can sometimes be answered unproblematically. The discovery of helium, which he described in considerable detail, is one such case [Kragh, 2009].

Having outlined the criteria of Rancke-Madsen and those of Kragh, Kuhn's distinction between predictable and unpredictable discoveries, and the project of Arabatzis and Gavroglu to describe the assimilation of argon into the scientific knowledge of the day, I will keep them in mind as I examine how argon, helium and neon came to be found and recognised as elements.

## Finding Argon, Helium and Neon

The story of the discovery of the noble gases has been told frequently over the last century and more. Accounts by participants include *The Gases of the Atmosphere* (1896–1915)[1] by William Ramsay (1852–1916) and *The Discovery of the Rare Gases* (1928) [Travers, 1928] by his assistant Morris Travers (1872–1961). Historical treatments by other scholars began soon afterwards [Weeks, 1932] and have continued to the present [Labinger, 2021]. In addition to these accounts, which cover most of the members of the noble gas group,

---

[1]The first edition appeared in 1896 in the wake of the discovery of argon. The second (1902) included neon, krypton and xenon. The third and fourth (1905 and 1915) also included material on the radioactive gases, now recognised as isotopes of radon.

I recommend the already mentioned accounts by Arabatzis and Gavroglu [2016] on argon and by Kragh [2009] on helium, as well as an earlier article of mine [Giunta, 2001], because they treat points relevant to the discussion below.[2] What follows is an account of the detection and characterisation of argon, helium and neon. That account is intended to provide a basis for the subsequent sections, which discuss the adequacy (in the cases of these elements) of the models of discovery summarised above.

## Argon

The detection and isolation of a new gas of the atmosphere now known as argon (for its inertness, from the Greek argon (ἀργός), lazy) grew out of a program of research into the densities of gaseous elements as a test of Prout's hypothesis (that all elements' atomic weights were multiples of those of hydrogen) [Arabatzis and Gavroglu, 2016]. After bringing to successful conclusions investigations of hydrogen and oxygen, Lord Rayleigh (John William Strutt, third Baron Rayleigh (1842–1919)) encountered a small but persistent anomaly while investigating nitrogen. Samples prepared from atmospheric air by removing other components were very slightly but consistently more dense than samples prepared by chemical reactions of nitrogenous compounds. Rayleigh, a physicist, published an appeal to chemists who might have been able to shed light on the anomaly [Rayleigh, 1892]. In April 1894, he read a paper to the Royal Society of London, describing the anomaly in more detail and experiments designed to troubleshoot its cause [Rayleigh, 1894].

Shortly after this presentation, Rayleigh and Ramsay (who had communicated with Rayleigh about preparing nitrogen from nitrogenous compounds) began to investigate the possibility that the relatively unreactive part of the atmosphere contained something besides nitrogen. They worked in parallel, by different methods, keeping each other apprised of developments.

[2] Additional discussions of the discovery of one or more of the noble gases can be found among the references of these sources.

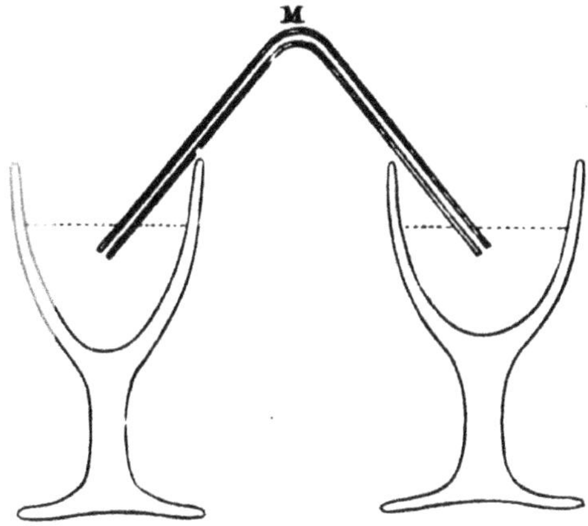


Figure 1: Apparatus Cavendish [1785, foldout facing p. 384] used to pass electricity through a sample of gas. The vessels contained mercury and the gas was trapped in the bent tube. In the public domain.

Rayleigh employed a method first reported by Henry Cavendish (1731–1810) [Cavendish, 1785], updated with late 19$^{th}$-century technology: using an electrical discharge to cause nitrogen to react with oxygen and removing the gaseous reaction products by reaction with an alkali. Cavendish had introduced the test gas and alkali ("soap lees") into the tube M (Figure 1). His vessels contained mercury, which (to use later terminology) served as electrodes. He introduced static electrical sparks into the system by bringing a charged sphere near enough to the mercury in one of the vessels for the charge to jump (spark) to the mercury while the mercury in the other vessel was connected to the ground. Rayleigh, in his first apparatus (Figure 2), kept his test gas in a test tube (A) inverted over a larger vessel containing an alkaline solution (B). He introduced electricity supplied from a battery of Grove cells through wires (D) encased in glass tubes (C). Reaction products would dissolve in the alkali; what was left in tube A was the unreactive gas. In later modifications, Rayleigh used the "public supply" of alternating current available at the Royal Institution as his source of electricity [Rayleigh and Ramsay, 1895].

Ramsay's apparatus is shown in Figure 3. He introduced his test gas, "atmospheric nitrogen", from reservoir F into a series of tubes.

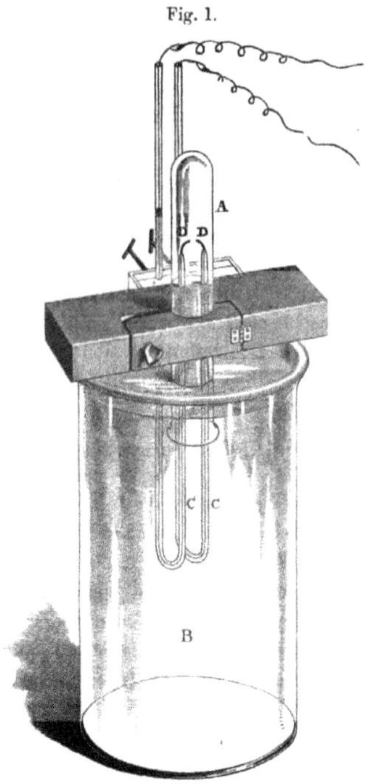


Figure 2: Rayleigh's apparatus for removing nitrogen from air by sparking [Rayleigh and Ramsay, 1895, p. 198]. In the public domain.

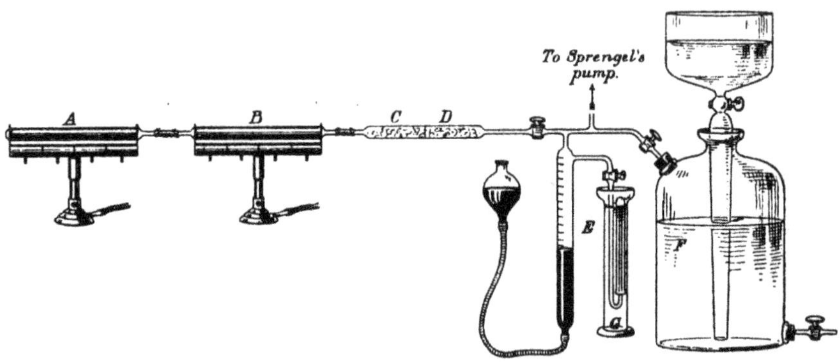


Figure 3: Ramsay's apparatus for removing nitrogen from air by reaction with hot magnesium [Rayleigh and Ramsay, 1895, p. 200]. In the public domain.

Red-hot magnesium in tube A reacted with nitrogen, producing solid magnesium nitride, which remained in the tube. Tube B contained hot copper oxide to react with any organic impurities or hydrogen. Tube CD contained phosphorus pentoxide to absorb stray moisture and soda lime to absorb stray carbon dioxide. Any residual gas was combined with the next batch of "atmospheric nitrogen", and the gas remaining after several cycles was the unreactive gas.

A letter from Ramsay to Rayleigh dated 4 August 1894 reported that he had isolated the unreactive gas. Rayleigh replied on 6 August that he believed he had also isolated the gas, and he suggested a joint publication. Ramsay, on 7 August, agreed that joint publication was appropriate.

Rayleigh made a brief presentation in both his and Ramsay's names to a hastily arranged session of the British Association for the Advancement of Science (the BA) at its annual meeting on 13 August in Oxford. From contemporary reports of the meeting [The Times, 1894; Nature, 1894; Chemical News, 1894a],[3] it is apparent that the presentation was a preliminary announcement of the discovery of a new gas in the atmosphere. Rayleigh described the methods by which the gas was isolated, reported a preliminary value for its density, and mentioned its spectrum. The chemical inactivity of the new gas was remarkable. The report in the *Chemical News* — but not those in *The Times* or *Nature* — explicitly supposed the gas to be elementary. The announcement in Oxford was certainly noticed. Lord Kelvin (William Thomson, 1824–1907) called it "the greatest scientific event of the past year" [Chemical News, 1894b, p. 301], while other scientists expressed scepticism [Chemical News, 1894c].

[3]The facts of the report but none of its content is recorded in the official conference proceedings. The contemporary reports in the general-interest, scientific, and chemical periodicals cited here list no authors. The report in the *Chemical News* was certainly written by its editor, William Crookes (1832–1919), an expert in spectroscopy.

Rayleigh and Ramsay continued to work on producing and characterising the new gas during the remainder of 1894, making no further public pronouncements on the subject. Ramsay presented a lengthy paper in both of their names at the Royal Society's meeting of 31 January 1895 [Rayleigh and Ramsay, 1895]. In addition to providing greater detail on aspects first mentioned in their preliminary announcement (i.e., on methods of isolation, density, and chemical inertness), they reported measurements that led them to conclude that the gas was monatomic. (The relevant measurement was of the ratio of specific heats under conditions of constant volume and constant pressure. The kinetic theory of gases permitted inferences to be drawn about the complexity of the "molecule" of the gas based on this ratio.) At the same meeting, William Crookes (1832–1919) read a paper on the visible emission spectra of argon, which were distinct from spectra of other gases or vapours he had examined [Crookes, 1895]. Karol Olszewski (1846–1915) reported on some thermophysical properties of argon, including its boiling and freezing points and critical constants [Olszewski, 1895].

Rayleigh and Ramsay [1895] concluded that the new gas was present in the atmosphere: it was isolated, not synthesised, by their experiments. The specific heat ratio led them to conclude that it was monatomic, and if it was monatomic, it must be an element or a mixture of elements. The definiteness of boiling and freezing points strongly suggested a pure substance rather than a mixture, but the authors did not rule out a mixture. Its extreme inertness and the distinctness of its spectra led them to conclude that it was not a previously known element. So, their best judgement on the nature of the new gas was that it was an element of atomic weight of about 40, to which they gave the name argon. They recognised that an element with the atomic weight and properties of argon would not fit in the periodic table:

> If argon be a single element, then there is reason to doubt whether the periodic classification of the elements is complete; whether, in fact, elements may not exist which cannot be fitted among those of which it is composed
>
> [Rayleigh and Ramsay, 1895, p. 236].

Where argon fit into the periodic system — if it fit at all — was a question much discussed in the scientific press after Rayleigh and Ramsay published these results [Giunta, 2001].

## *Helium*

Accounts of the discovery of helium usually begin with the observation of a spectral line in solar prominences in 1868. The line was first observed on 18 August by Pierre-Jules-César Janssen (1824–1907), a French astronomer in India observing a solar eclipse. On 20 October of the same year, Joseph Norman Lockyer (1836–1920) also observed that line, among others. Both sent notices to the *Académie des Sciences* in Paris; both notices arrived on the same day (26 October).[4] The line was close to but not coincident with the lines in the solar spectrum labelled D by Joseph von Fraunhofer (1787–1826). Figure 4 displays a solar absorbance spectrum (the white band with dark lines). Underneath it are the labels Fraunhofer had assigned to various lines and the elements responsible for many of them. Well before the time of the observations of Janssen and Lockyer, Fraunhofer's D feature was seen to consist of two closely spaced lines labelled $D_1$ and $D_2$. Above the absorbance spectrum in Figure 4 is an emission spectrum from a solar prominence. Just to the left of the two D lines from sodium can be seen the newly

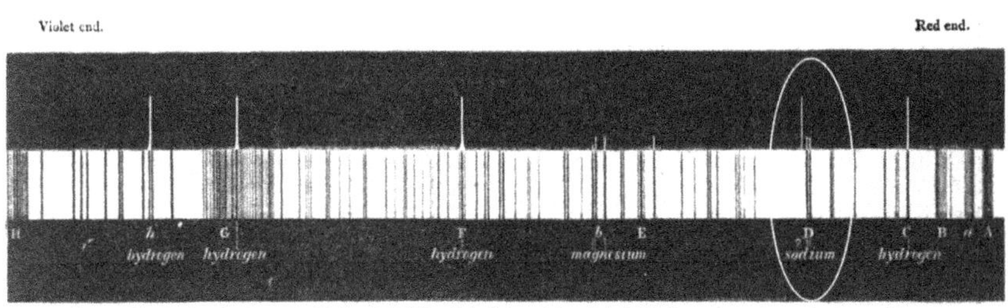


FIG. 1.—Showing the solar spectrum, with the principal Fraunhofer lines, and above it the bright-line spectrum of a prominence containing magnesium, sodium, and iron vapour at its base.

Figure 4: Spectra presented by Lockyer in the first issue of *Nature* [Lockyer, 1869, p. 14]. The oval around the D feature was added by the present author. In the public domain.

[4] This section is indebted to Kragh [2009] for the story of helium before 1895.

reported line, which came to be denoted as $D_3$. At first, $D_3$ was only seen in sunlight and later in starlight as well; it did not belong to the spectrum of any known element.

The name helium made its public debut in the 1871 inaugural address to the BA by William Thomson [1871]. He said that Edward Frankland (1825–1899) and Lockyer believed the spectral line near D to belong to a new substance "which they propose to call Helium". (The name helium refers to the sun, helios (*ἥλιος* in Greek).) Lockyer [1896] coined the term while he was working with Frankland on the possibility that the $D_3$ line belonged to hydrogen under extreme conditions. Neither Lockyer nor Frankland ever formally proposed a new element or even a new substance on the basis of the $D_3$ line, but the name helium began to appear in the scientific literature, sometimes attached to a hypothetical element.[5]

One report of helium on earth appeared before 1895. In 1881, Luigi Palmieri (1807–1896) observed the $D_3$ line from an amorphous specimen collected from the site of a recent eruption of Vesuvius. His very brief preliminary note was not followed by additional detail.

The day after the detailed presentation on argon at the Royal Society, Henry Miers (1858–1942), Keeper of the Collection of Minerals in the British Museum, wrote to Ramsay to suggest that looking into natural uranate minerals might be worthwhile, for they were reported to contain nitrogen. He gave Ramsay a reference to a report by the American mineralogical chemist William Hillebrand (1853–1925) [Hillebrand, 1890]. Ramsay took the suggestion, hoping that he might find a compound of argon. He purchased some cleveite and gave his assistant, Donald John Matthews (1873–1956), the task of treating the mineral and collecting the released gas. Ramsay and Crookes each examined the spectrum of the gas, which displayed a bright yellow line: they independently concluded that

[5]See, for example, John Draper (1811–1882), in his inaugural address as first president of the American Chemical Society: "I often look at the bright yellow ray emitted from the chromosphere of the sun, by that unknown element, Helium, as the astronomers have ventured to call it" [Draper, 1876, p. 209].

the gas from cleveite contained a new gas [Travers, 1956]. Crookes identified its bright yellow line as the $D_3$ of helium. Ramsay immediately informed Marcellin Berthelot (1827–1907), permanent secretary of the French *Académie des Sciences*, who communicated the report at the 25 March session of the *Académie* [Berthelot, 1895b]. Not long afterward, the Swedish chemist Per Teodor Cleve (1840–1905), for whom cleveite is named, reported that his student Nils Abraham Langlet (1868–1936) had also found helium, but not argon, in cleveite [Kragh, 2009].

Now that a substance linked to the $D_3$ line and the name helium had been "run to earth", questions remained about its characteristics. Several weeks later, Ramsay began to report preliminary measurements of helium's physical properties. Its density was no more than 3.89 (compared to oxygen 16), but the sample was not entirely free of nitrogen.[6] He reported a ratio of specific heats that he had reason to believe was not accurate (for it was higher than the theoretical limit from the kinetic theory of gases); from this, he inferred that the ratio was high enough to indicate that helium is monatomic [Ramsay, 1895]. Langlet found its density to be 2.02 times that of hydrogen, lower than Ramsay's reported value [Cleve, 1895]. On 20 June 1895, a paper by Ramsay, John Norman Collie (1859–1942) and Travers on the sources and properties of helium was presented to the Chemical Society. They reported that helium was similar to argon in its inertness and its specific heat ratio of 1.66 and that its density was 2.13 [Ramsay *et al.*, 1895].

## *Neon*

The place of argon in the periodic table will be discussed below, but some of Ramsay's proposals demand attention here as part of the prehistory of the discovery of neon. In his monograph *The Gases of*

[6] A 21st-century reader who knows that this value is close to 2.00 might wonder how a value so far from the accurate value could be reported with such precision. The data given in the paper suggest that the density measurement was as precise as was reported, but the sample in question was not pure helium, as Ramsay acknowledged at the time.

THE PERIODIC SYSTEM.

| | | | |
|---|---|---|---|
| | | Hydrogen . 1·01 | Helium . . 4·2 |
| Nitrogen . 14·0 | Oxygen . . 16·0 | Fluorine . 19·0 | ? |
| Phosphorus 31·0 | Sulphur . . 32·1 | Chlorine . 35·5 | Argon . . 39·9 |
| Arsenic . 75·1 | Selenium . 79·0 | Bromine . 80·0 | ? |
| Antimony . 120·3 | Tellurium . 126·3 | Iodine . . 126·9 | ? |
| Erbium . . 166·0 | ? . . . . 167·0 | ? . . . . 169·0 | ? |
| Bismuth . 208·1 | ? . . . . 214·0 | ? . . . . 219·0 | ? |

Figure 5: Fragment of Ramsay's periodic table in *Gases of the Atmosphere* [1896, p. 219]. In the public domain.

*the Atmosphere* [Ramsay, 1896, p. 219], Ramsay placed argon between chlorine and potassium, that is, not in order by atomic weight. The periodic table in this book shows helium and argon in a new group, with question marks holding the places of members as yet not discovered (Figure 5).

Ramsay's presidential address to the chemical section of the BA in 1897 made a more explicit and public prediction of a yet unknown member of the helium-argon group of elements. The "undiscovered gas" of the lecture's title was a monatomic gas of atomic weight about 20 "as indifferent to union with other elements" [Ramsay, 1897, p. 92] as argon and helium. He and Travers were looking for this unknown gas, starting their search in minerals, which is where helium had been found.

Travers [1956, pp. 170–171] wrote that the Irish physicist George Johnstone Stoney (1826–1911) gave Ramsay the idea to look for the undiscovered gas in the atmosphere. Gases as light as hydrogen and helium could reach speeds in Earth's upper atmosphere that permitted them to escape Earth's gravity, but water vapour was heavy enough not to do so appreciably — and Ramsay's undiscovered gas was slightly heavier than water vapour. In 1898, large quantities of liquid air were made available to Ramsay and his co-workers, and they thought to look there for congeners of argon. In their first experiment, they let most of a sample of liquid air evaporate away and retained the last-boiling fraction to look for a residue after reacting away oxygen and nitrogen. At the Royal Society meeting of 9 June 1898, they described evidence (from spectral, density, and specific heat measurements) that led them to conclude that the residue was a mixture of argon with a heavier, as yet unknown, monatomic (and therefore elementary) gas. They proposed the name "krypton" (meaning hidden in Greek (κρυπτόν)) for it. The last paragraph of the report announced their intention to look for a gas lighter than nitrogen in atmospheric air [Ramsay and Travers, 1898a].

One week later (16 June), they were back to announce more companions of argon [Ramsay and Travers, 1898b]. They had been accumulating argon from atmospheric air for some time in preparation for the experiment. Now they used liquid air to cool and liquefy most of their argon supply, and they collected the vapour that remained over the liquid argon. They examined the spectrum and density of that vapour. The spectrum convinced them that the vapour contained a previously unknown component, for which they proposed the name neon (meaning young or new, from the Greek neon (νέον)). The density of the vapour was lower than that of argon, which was consistent with their expectation of finding a lighter congener of argon. The paper did not contain a claim of elementary status for neon. It did, however, contain a claim for another monatomic element, "metargon", having the same density as argon, but a different spectrum.

A little over two years passed before Ramsay and Travers published a detailed description of the purification process and of several properties of argon and its companions, as well as a retraction of the claim of metargon [Ramsay and Travers, 1901]. Densities of reasonably pure samples of helium, neon, argon, krypton, and xenon and their atomic weights (assuming, based on specific heat ratios, monatomic gases) are shown in Table 1. The authors claimed that the gases "form a group by themselves" in the periodic table. They added data points from these gases to plots of refraction equivalents and atomic volumes that had previously illustrated the periodicity of properties. The new plots, including the inactive gases, still illustrated the periodic law (Figure 6).

Table 1: Densities (relative to oxygen = 16) and atomic weights of the inactive gases [Ramsay and Travers, 1901].

| Gas | Helium | Neon | Argon | Krypton | Xenon |
|---|---|---|---|---|---|
| Density | 1.98 | 9.96 | 19.96 | 40.78 | 64.0 |
| Atomic weight[7] | 4 | 20 | 40 | 82 | 128 |

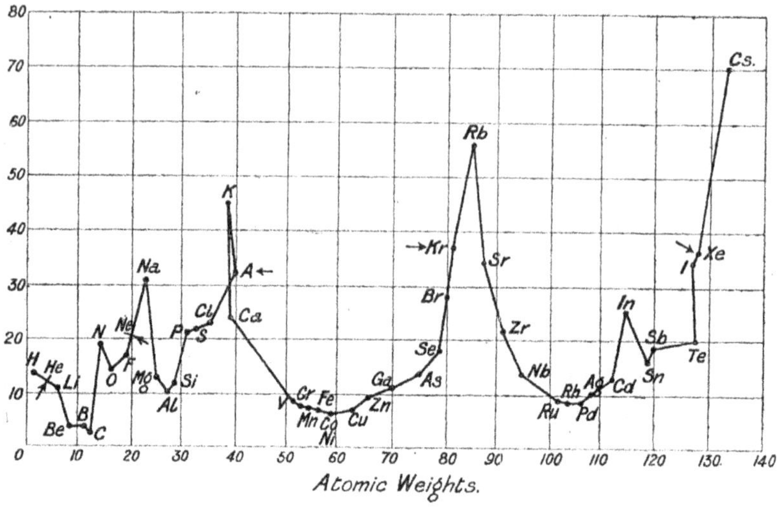


Figure 6: Plot of atomic volume vs. atomic weight illustrating the periodicity of this property [Ramsay and Travers, 1901, p. 89]. Arrows point to the data points of the inactive gases. In the public domain.

## Applying the Criteria of Rancke-Madsen

According to the criteria proposed by Rancke-Madsen [1975], the discoverers of argon, helium and neon and the dates of effective discovery were:

[7]The densities are reported as experimental numbers consistent with the precision of the experiments but the atomic weights (which depend on the densities) are reported only to the nearest integer. It was not unusual for chemists throughout the 19th century to report atomic weights as integers. Considering that Rayleigh found argon in the course of investigating Prout's hypothesis, the treatment of atomic weights as if they were integers by his erstwhile collaborator is puzzling.

- for argon, Rayleigh and Ramsay on 13 August 1894 (announcement at the BA meeting).
- for helium, Ramsay (and possibly Matthews and Crookes?) on 25 March 1895 (communication of Ramsay's preliminary report to the French *Académie*).
- for neon, Ramsay and Travers on 16 June 1898 (Royal Society meeting at which they made their preliminary announcement).

These statements do not adhere to the common usage of "discovery" or indeed to the specific usage of the discoverers,[8] although they do conform to one sense of "discover", namely, to disclose or reveal. Rancke-Madsen was surely correct to include disclosure among his criteria for the effective discovery of an element, but misguided, I think, in dating a discovery to what he considered its final act.[9] One can demand that an observation or experiment not be considered a discovery until it is revealed in public yet still speak of the observational or experimental work as the discovery and the disclosure as the announcement of the discovery.

Recall that Rancke-Madsen's criteria, in effect, require the discoverer to find the substance that is or contains a new element and to recognise it as new but not necessarily identify it as an element. The investigators named in the above list had collected convincing evidence that they had found novel species on or before the listed dates, but they had not yet necessarily collected convincing evidence that what they had found were elements.

Lockyer may well have thought that the $D_3$ line implied the existence of a novel substance (later recognised as an element), but in the absence of a tangible substance, the basis for such an inference was insufficient both logically and socially.[10] Until helium was found on

[8] Ramsay wrote "announcement of the discovery" (of argon) [Ramsay, 1896, p. 164] and Ramsay and Travers wrote that they had "announced the discovery" (of xenon) [Ramsay and Travers, 1901], making a distinction between the discovery and its communication.

[9] See also Annette Lykknes' contribution to this volume.

[10] See also Brigitte Van Tiggelen's contribution to this volume.

earth, its status as an element was based on evidence no stronger than that for the spurious element coronium, and no consensus of scientists accepted it [Trimble, 2021]. Palmieri's report [1881] of terrestrial helium at Vesuvius was an isolated report, neither substantiated by the researcher nor "noticed" by the wider community. Hillebrand [1890] surely collected helium, but he made no claim that he had found an unknown species. None of them, therefore, made an effective discovery of helium.[11]

## Discovery of Argon, Helium, and Neon as Extended Processes

In this section, I examine the discoveries of argon, helium and neon in light of the characterisations by Kuhn [1962] and by Arabatzis and Gavroglu [2016] of discovery as an extended process.

### *Analysis from Kuhn's Perspective*

Kuhn [1962] distinguished between predictable and unpredictable discoveries. Only neon, of the elements under consideration here, was a predictable (indeed, predicted) discovery. When Ramsay started looking for it, he had a good idea of what he was seeking: an unreactive monatomic gas whose atomic weight was about 20. He also had a good idea of how to test candidates. Kuhn would expect questions of when and where the discovery occurred to be answerable. Neon was discovered in Ramsay's laboratory at University College London. The spectrum of a sample of gas was taken, and the density and specific heat of the gas were measured on Sunday, 12 June 1898 [Travers, 1956]; Ramsay and Travers [1898b] reported these measurements to the Royal Society on 16 June as the basis for their claim of a new constituent of the atmosphere they called neon.

So, was neon discovered on 12 June 1898? One may conclude that the data collected that day were sufficient to permit Ramsay and

[11] See also Kragh's contribution to this volume.

Travers to assert that their sample of gas contained a new element (even though they made no such assertion on 16 June). Kuhn noted that statements about unpredictable discoveries, which imply that they occurred in "an instant in time", are misleading, and I maintain that they are equally misleading, even in the case of this predicted discovery. The sample of gas tested on 12 June was prepared over the previous two days by chilling and evaporating a large supply of argon. A large supply of argon was accumulated, starting early in 1898 in preparation for this experiment. The search for neon began by other means during the previous year [Travers, 1956]. The data collected on 12 June 1898 were not sufficient to place neon in the periodic table. If placement in the periodic table was necessary, as Arabatzis and Gavroglu argued in the case of argon, the atomic weight of neon was required; for that, a reasonably pure sample of neon was needed, meaning that more neon had to be collected and purified.

The plurality of defensible beginning and end points of the discovery of neon is not intended to suggest that the question "When was neon discovered?" does not have an answer: an extended period of time is a perfectly valid answer, just as it is for questions such as "When did Michelangelo paint the Sistine Chapel?" My point is that the answer to the "when" question is rarely "an instant in time", even for predictable discoveries. The "when" question is not objectionable in itself. What is objectionable is an expectation that its answer be an instant — a "eureka moment" — whether or not the discovery is predictable.

As has already been noted, the questions that the individualist model of discovery asks can be answered unproblematically for the discovery of the element helium [Kragh, 2009]. This was an unpredictable discovery, but its temporal development was not very different from that of neon. Once Ramsay accepted the suggestion to investigate uranate minerals, it took time to acquire appropriate samples and for his assistant (Matthews) to collect gas from the mineral before the spectrum of the gas was recognised as that of helium. This recognition resembles a "eureka moment" in some accounts [Travers, 1956], for good reason: a connection was suddenly made between a mysterious solar emission and a manipulable terrestrial material. The identification of the gas with the hitherto enigmatic helium was suggestive that Ramsay had found an element; clearly, it was not a known element under extreme conditions (as Lockyer and Frankland had wondered), but, to determine that the gas was monatomic and, therefore, elementary, required preparing more of the gas and taking additional measurements.

The discovery of the element argon was clearly an extended process, even if one starts the timer in the spring of 1894 when Rayleigh and Ramsay began to remove nitrogen (and the other known components) from atmospheric air. Kuhn's assertion that unpredictable discoveries began with anomalies and that "making the anomaly behave" was an important phase certainly applies to this case: Rayleigh had spent much time documenting an anomaly in his measurements of the density of nitrogen and trying to track down its source. By August 1894, Rayleigh and Ramsay had evidence sufficient to support the claim that they had found a new gas in the atmosphere, and its spectrum and inertness were certainly suggestive that it contained a new element or was a new element. By the end of January 1895, they had collected enough evidence to demonstrate by the criteria of the time that the new gas was an element.

## *Analysis from the Perspective of Arabatzis and Gavroglu*

Arabatzis and Gavroglu [2016] argued that the discovery of the element argon was not complete until argon was integrated into the science of the time, altering thereby what was accepted knowledge; in doing so, they followed Kuhn. I maintain that what altered the accepted science of the day was not part of the discovery of the element argon but rather its aftermath; in effect, I assert that a boundary can be drawn between the discovery of argon as an element and the adjustments to accepted knowledge brought about by the unexpected properties of the noble gases. Arabatzis and Gavroglu proposed that two particular pieces of chemical knowledge prevented the acceptance of argon as an element, namely, the expectation that chemical elements would have a place in the periodic system and that they would combine with other elements. Argon, they contended, was admitted as an element once it found a place in the periodic system and when chemical reactivity was dropped as a necessary characteristic of elements — this resolution was facilitated by finding gases similar to argon. I argue below that the acceptance of a discovery is separate from the act of discovery; moreover, I doubt that the issue of chemical reactivity actually delayed the acceptance of argon as an element, and I am not persuaded that its place in the periodic system did so either.

Arabatzis and Gavroglu presented one quotation supporting the contention that the chemical inactivity of argon delayed its recognition as an element: the Irish physicist George Francis FitzGerald (1851–1901) advised Ramsay that his fellow chemists would not accept an element that had no chemical affinity. Among many comments about argon published in the *Chemical News* after the initial announcement of the gas, I found only one that opined that the term element ought not to apply to an atom incapable of combining with other atoms [Sedgwick, 1895].

I do not doubt that chemists in 1894 assumed that all chemical elements were capable of combining with other elements. However, if reactivity was part of their working definition of an element, it was a tacit part. Chemical textbooks and treatises generally defined elements as bodies that could not be decomposed into simpler bodies. I compared definitions of elements from books by three prominent chemists in editions before the discovery of the noble gases to later editions that included those gases [Mendeléeff, 1891/1905; Remsen, 1890/1907; Tilden, 1888/1905]. The property of reactivity did not appear in the explicit definition of an element either before or after the discovery of the noble gases.[12] Furthermore, although the chemical inactivity of argon was marked, its complete inactivity was at least an open question for some time. Scientists sought, and in some cases reported, compounds of argon [Parker, 1897]; the most

[12]William A. Tilden's (1842–1926) book defined elements in terms of microscopic entities (atoms) in 1888 and earlier editions. In 1905, the definition was macroscopic in terms of substances that could not be decomposed into simpler substances.

prominent among them was Berthelot [1895a]. I cannot rule out reluctance to accept an inert element as an unstated reason behind interpretations of argon as something else, such as an allotrope of nitrogen. However, there seems to be little evidence that its inertness delayed its acceptance as an element.

Could argon have been considered a new element before it was placed in the periodic system? Certainly, some chemists thought that it could — including Dmitrii Ivanovich Mendeleev. Mendeleev's reaction to the announcement of argon, reported in *Nature* in spring 1895, accepts that argon is a simple substance, not a mixture or a compound. To be sure, he was sceptical of the conclusion that argon was a monatomic gas, noting, "If we admit that the molecule of argon contains but one atom, there is no room for it in the periodic system." His favoured interpretation of Rayleigh and Ramsay's research was that argon was an allotrope of nitrogen, $N_3$ [Mendeléeff, 1895]. One can describe Mendeleev's opinion using later terminology by saying that he accepted argon as a previously unknown elementary substance, most likely a previously unknown form of a previously recognised element.

Rayleigh and Ramsay [1895, p. 236] speculated whether "elements may not exist which cannot be fitted among those of which it [the periodic classification] is composed". Others were willing to grant argon elemental status even while trying to place it into the periodic table. The English chemist John Hall Gladstone (1827–1902), for example, asserted that unless and until argon can be shown to be convertible into another element, "it has the full right to take rank as a new element" [Gladstone, 1895, p. 389]. The Irish chemist James Emerson Reynolds (1844–1920) wrote a letter to *Nature* he thought would interest readers "who are actively discussing the probable position of 'argon', on the assumption that this remarkable substance is an element" [Reynolds, 1895, p. 486]. This statement does not illustrate that Reynolds believed argon to be an element, but it does indicate his opinion that its place in the periodic table is a question separate from its elementary character.

The properties of argon announced in early 1895 were certainly difficult to integrate into the accepted science of the day (as I have described elsewhere [Giunta, 2001]). I would not be surprised if no consensus about the elementary status of argon developed before it was placed in the periodic system. But I do not agree that the discovery of the element argon extended until its place in the periodic system was fixed.

Consensus among the scientific community is a prerequisite to the addition of a concept or object to the edifice of scientific knowledge.[13] To date in real-time when such a consensus is reached is difficult, if not impossible, and even to date it in retrospect is challenging. In any event, I contend that although consensus is needed before a claim ought to be described as a discovery, the date of the discovery is the date of the investigation, not of the achievement of consensus. Thus, we can speak of "acceptance of a discovery", just as we can of "announcement of a discovery", as something akin to certification of a past event: the certification is necessary but not a part of the event.

## A Journalistic Model of Discovery

Based on the analysis and arguments made above, I suggest that the discovery of a chemical element can be well described by answering the journalistic questions associated with the individualist model of scientific discovery:

> What was discovered?
> Who made the discovery?
> When was the discovery made?
> Where was the discovery made [Arabatzis and Gavroglu, 2016, p. 204]?

[13] See Oreskes [2019] for an interesting and wide-ranging discussion of how the social character of science lends reliability to the claims of science.

Along with two additional basic questions, also from journalism[14]

> How was the discovery made?
> Why was the discovery made?

I call a model of scientific discovery characterised by thorough answers to these questions a journalistic model. I consider the term "journalistic model" to be a convenient shorthand, although journalists have no monopoly on this set of questions.

Let me make clear that I do not promote the individualist model; Kuhn, Kragh, and Arabatzis and Gavroglu were right to critique it. Its main problem, though, is not its questions but assumptions about their answers. As Arabatzis and Gavroglu posed those questions about the discovery of argon:

> There are difficulties in attributing the discovery of argon to any particular scientist (Cavendish? Rayleigh? Ramsay?); to pinpoint its date (1892? 1895? Later?); to specify its location (which particular laboratory?); and, more importantly, to identify the object of discovery (what was argon? Or, to put it less provocatively, how did chemists and physicists conceptualize the new gas that they had detected?).
>
> [Arabatzis and Gavroglu, 2016, p. 205]

The answer to "where" need not be a single laboratory. As discussed above, the answer to "when" need not be a single day or a "eureka moment". Similarly, "who" need not be a single individual; it can be a group of collaborators or independent researchers. The possibility of multiple candidates does not preclude listing some as discoverers and not others based on informed judgement. The "who" question is complicated by changing standards for recognising the contributions of collaborators, assistants and technicians.

[14] See Hicks *et al.* [2008, p. 13] for an example of a journalism textbook treating these six basic questions.

The first question on the list, "What was discovered?" is open-ended for scientific discoveries in general but more limited for the discovery of elements. Kuhn [1962, pp. 761–762] noted that discovery requires "recognizing both *that* something is and *what* it is" and "there are no benchmarks to inform either the scientist or the historian when the job of discovery has been done". To answer the "what" question for one of the elements under consideration here, helium, illustrates the lack of benchmarks. What is helium? Ramsay and Travers found that it is a monatomic gas of atomic weight 4, whose spectrum had been observed before the gas was isolated. Today, we might add that it is element 2, produced by nuclear fusion in stars and by alpha radioactive decay; at very low temperatures, it can be made superfluid, and at very high pressures, it can be forced into combination with sodium [Dong *et al.*, 2017]. Surely, discovering what helium is has not yet ended.

Discovering an element, though, is not so open-ended. Kragh's [2019] model for discovering elements does have benchmarks: finding (or making) a novel substance and recognising it as an element. As a conceptual discovery, recognising an element depends on the concept of an element and the methods accepted at the time. In the 1890s, an element was a substance not decomposable into simpler substances. Analysis of visible spectra has been accepted since the 1860s as strong evidence of novelty. As experimental techniques and the concept of elements changed, taking into account, for example, the atomic number, so did the evidence and methods for recognising new elements [Orna and Fontani, 2021]. Twenty-first-century scientists clearly believe that criteria can be established for evaluating claims of the discovery of new elements, as evidenced by the work of international bodies to develop, apply,[15] and revise such criteria [Hofmann *et al.*, 2020]. These criteria are consistent with a journalistic model; however, having a narrower scope and purpose than historians and philosophers attempting to understand the nature of scientific discovery, they neglect the "why" question while placing a lot of emphasis on "how".

[15] The most recently recognized element claims were evaluated using an earlier set of criteria, namely those published in [IUPAC and IUPAP, 1991].

The questions of "how" and "why" were not listed among the objectionable questions of the individualist model, perhaps because they are less amenable to the short and simple answers. "How" it seems to me to be particularly relevant to scientists evaluating the validity of claims of element discovery: how were the experiments done, and what were the results? Even short answers to the "how" question, e.g., by electrolysis, confirmed by visible spectroscopy, by distillation from air, etc., can illuminate trends in methods of element discovery. The "why" question is perhaps the most open-ended of all. It is amenable to answers that provide context from internalist and externalist perspectives (that is, motivations that derive from within science and outside it, respectively).

***

I close by sketching answers to the journalistic questions for the discovery of the elements considered here, applying Kragh's criteria [2019] and beginning with answers proposed for argon by Arabatzis and Gavroglu [2016]. I hope that these answers show that a journalistic model provides a useful framework for understanding the discoveries in question. The journalistic model leaves room for drawing boundaries based on criteria, such as describing certain experiments as part of the discovery of argon and others as part of earlier or later investigations. It permits open-ended answers to apparently simple questions, such as describing the actions and contributions of several possible discoverers. And it allows much latitude for contextualisation through the "why" question — which I address only scantily.

Who discovered argon? Cavendish probably concentrated a small amount of argon, but he did not try to characterise the residue of his experiments other than by volume ("not more than 1/120 part of the whole") [Cavendish, 1785, p. 382]. He did not claim to have discovered a new air, let alone a new element. So, Cavendish did not discover the element argon. Rayleigh and Ramsay worked separately to isolate the gas and then to collect and characterise it, reporting their results jointly. They qualify as discoverers. Do others also qualify? For example, Crookes's [1895] paper provided important spectral evidence that argon was novel — relevant to the claim that argon was an element. Olszewski [1895] also presented a paper on the same occasion: his results suggested that argon was a pure substance but had little bearing on its status as an element. Rayleigh and Ramsay also acknowledged the help of four of Ramsay's assistants: Percy Williams (1873–1942), who was singled out, Alexander Mitchell Kellas (1868–1921), Matthews (mentioned earlier) and Gordon (about whom I have no further information). A thorough answer to the "who" question would credit these people for their roles but not, I believe, for discovering argon.

When? Certainly not as early as 1892, when Rayleigh asked for help from chemists in making his anomaly behave. Rayleigh and Ramsay set about trying to isolate the heavier component from "atmospheric nitrogen" in spring 1894. In early August, each succeeded. After that, they worked on characterising the gas. They presented data meeting the criteria of having found a new element on 31 January 1895. So, they discovered argon in the course of an investigation that ran from spring 1894 to January 1895. Laboratory notebooks could probably furnish more exact start and end dates.

Where? Argon was discovered in Rayleigh's laboratory at his home, Terling Place, northeast of London, in Ramsay's laboratory at University College London, and at the Royal Institution, where Rayleigh prepared larger quantities of argon useful in establishing its elementary nature.

What was discovered? Rayleigh and Ramsay could answer the question by pointing to a sealed tube; its contents were what they discovered. In their investigation, they learned that what they had discovered was a monatomic gas, different from any gas previously reported; therefore, it was a new element. By the time they knew that they had a new element, they also knew its atomic weight and that it was very inactive chemically. But they did not yet know how or even whether it fit into the periodic system of the elements. That story is more important than the discovery of the element argon, but it is another story.

How? Both researchers isolated argon by removing other components of atmospheric air, each using a different chemical reaction to remove nitrogen. "How" is described more minutely above in this chapter and in still greater detail in Rayleigh and Ramsay [1895].

Why? Rayleigh's program to test Prout's hypothesis by measuring the densities of elementary gases appears to be the chief internalist reason for the line of inquiry that eventually led to argon.

For helium, I begin with the "how" question, for its answer can help answer the others. This gas was collected and purified from a sample of the mineral cleveite. Visible spectroscopy established the novelty of the gas. Specific heat measurements implied that it was monatomic. What was discovered? A new monatomic gas and, therefore, a new element. Who discovered helium? Kragh [2009] gave a reasoned discussion of this question, concluding that Ramsay did so, aided by his assistants. Matthews collected the gas from the mineral and purified it [Travers, 1956, pp. 139–140]. Crookes identified the mysterious solar line in the spectrum of the new gas. Where was it discovered? Except for the spectral analysis done by Crookes at his home laboratory in London, the research was carried out in Ramsay's laboratory at University College London. When? I concur with Kragh [2009] in citing the key moment to be the identification of the $D_3$ line in the spectrum of the gas, but I would date the discovery to a more extended period starting when Matthews began preparing the gas (a date I do not know, but before 14 March 1895) and ending when the gas was shown to be monatomic (by 25 April 1895, as Ramsay wrote to his wife [Travers, 1956, p. 150]). Why was helium discovered? Its discovery is a tale of many contingencies: the external circumstance of Miers's note to Ramsay and Ramsay's hope to find a compound of argon (an internal motivation) are among them.

How was neon discovered? It was separated from liquid air. Visible spectroscopy established the novelty of the gas, and specific heat measurements showed that it was monatomic. What was discovered? A new monatomic gas, and therefore, a new element. Who discovered neon? Ramsay and Travers, with some help from other assistants. Where was it discovered? Once again, in Ramsay's laboratory at University College London. When? The key date was 12 June 1898, when its visible spectrum and specific heat were measured. I would date the discovery to an only slightly more extended period beginning on 10 June 1898, with the chilling and evaporation of argon for the purpose of looking for a lighter monatomic gas. Why was neon discovered? Again, there is much relevant background, including Ramsay's prediction of this "undiscovered gas", Stoney's suggestion that the atmosphere was a good place to look for it, and the availability of liquid air [Travers, 1956].

## References

Arabatzis, T. and Gavroglu, K. (2016). From discrepancy to discovery: How argon became an element. In *The Philosophy of Historical Case Studies* [Boston Studies in the Philosophy and History of Science 319], eds. Sauer, T. and Scholl, R. (Springer, Switzerland), pp. 203–222.

Berthelot, M. (1895a). Essais pour faire entrer l'argon en combinaison chimique. *Comptes rendus de l'Académie des Sciences*, **120**, pp. 581–585.

Berthelot, M. (1895b). Nouvelles recherches de M. Ramsay sur l'argon et sur l'hélium. *Comptes rendus de l'Académie des Sciences*, **120**, pp. 660–662.

Cavendish, H. (1785). Experiments on air. *Philosophical Transactions of the Royal Society*, 75, pp. 372–384.

Chemical News. (1894a). A supposed new gaseous element in the atmosphere. *Chemical News*, 70 (24 August), p. 87.

Chemical News. (1894b). The Royal Society Anniversary Meeting, November 30$^{th}$, 1894. *Chemical News*, 70, pp. 288–292, 303–306 (at 291–292).

Chemical News. (1894c). Chemical Society, Ordinary Meeting, December 6$^{th}$, 1894. *Chemical News*, 70, pp. 300–303.

Cleve, P. (1895). Sur la densité de l'hélium. *Comptes rendus de l'Académie des Sciences*, **120**, p. 1212.

Crookes, W. (1895). On the spectra of argon. *Philosophical Transactions of the Royal Society (London) A*, **186**, pp. 243–251.

Dong, X., Oganov, A., Goncharov, A. *et al.* (2017). A stable compound of helium and sodium at high pressure. *Nature Chemistry*, **9**, pp. 440–445. https://doi.org/10.1038/nchem.2716

Draper, J. W. (1876). Science in America. *American Chemist*, **7**, pp. 205–213.

Giunta, C. J. (2001). Argon and the periodic system: The piece that would not fit. *Foundations of Chemistry*, **3**, pp. 105–128.

Gladstone, J. H. (1895). Argon. *Nature*, **51**, pp. 389–390.

Hicks, W. *et al.* (2008). *Writing for Journalists* (Routledge, New York).

Hillebrand, W. F. (1890). Occurrence of nitrogen in uraninite and composition of uraninite in general. *American Journal of Science*, **40**, pp. 384–394.

Hofmann, S., Dmitriev, S. N., Fahlander, C. *et al.* (2020). On the discovery of new elements: Report of the 2017 Joint Working Group of IUPAC and IUPAP. *Pure and Applied Chemistry*, **92**, pp. 1387–1446.

IUPAC and IUPAP. (1991). Criteria that must be satisfied for the discovery of a new chemical element to be recognized. *Pure and Applied Chemistry*, **63**, pp. 879–886.

Kragh, H. (2009). The Solar element: A reconsideration of helium's early history. *Annals of Science*, **66**, pp. 157–182.

Kragh, H. (2019). Controversial elements: Priority disputes and the discovery of chemical elements. *Substantia*, **3**(Suppl. 5), pp. 79–90. doi:10.13128/Substantia-740

Kuhn, T. S. (1962). Historical structure of scientific discovery. *Science*, **136**, pp. 760–764.

Labinger, J. A. (2021). The history (and pre-history) of the discovery and chemistry of the noble gases. In *150 Years of the Periodic Table: A Commemorative Symposium* [Perspectives on the History of Chemistry], eds. Giunta, C. J., Mainz, V. V. and Girolami, G. S. (Springer, Switzerland), pp. 303–327.

Lockyer, J. N. (1869). The recent total eclipse of the sun. *Nature*, **1**, pp. 14–15.

Lockyer, J. N. (1896). The story of helium. *Nature*, **53**, pp. 319–322, 342–346.

Lord Rayleigh (1892). Density of nitrogen. *Nature*, **46**, pp. 512–513.

Lord Rayleigh (1894). On an anomaly encountered in determinations of the density of nitrogen gas. *Proceedings of the Royal Society (London)*, **55**, pp. 340–344.

Lord Rayleigh and Ramsay, W. (1895). Argon, a new constituent of the atmosphere. *Philosophical Transactions of the Royal Society (London) A*, **186**, pp. 187–241.

Mendeléeff, D. (1891). *The Principles of Chemistry*, trans. Kamensky, G. (Longmans, Green, & Co., London), Vol. 1; 3rd ed. 1905.

Mendeléeff, D. (1895). Professor Mendeléeff on argon. *Nature*, **51**, p. 543.

Nature. (1894). The British Association. *Nature*, **50** (16 August), pp. 369–392 (at 370).

Olszewski, K. (1895). The liquefaction and solidification of argon. *Philosophical Transactions of the Royal Society (London) A*, **186**, pp. 253–257.

Oreskes, N. (2019). *Why Trust Science*? (Princeton University Press, Princeton, NJ).

Orna, M. V. and Fontani, M. (2021). Discovery of three elements predicted by Mendeleev's Table: Gallium, scandium, and germanium. In *150 Years of the Periodic Table: A Commemorative Symposium* [Perspectives on the History of Chemistry], eds. Giunta, C. J., Mainz, V. V. and Girolami, G. S. (Springer, Switzerland), pp. 227–257.

Palmieri, L. (1881). Della riga dell'helium apparsa in una recente sublimazione vesuviana. *Rendiconto dell'Accademia delle scienze fisiche e matematiche (Napoli)*, **20**, p. 233.

Parker, C. L. (1897). Our present knowledge of argon, with a partial bibliography. *Journal of the American Chemical Society*, **19**, pp. 124–138.

Ramsay, W. (1895). Helium, a gaseous constituent of certain minerals, part I. *Proceedings of the Royal Society (London)*, **58**, pp. 81–89.

Ramsay, W. (1896). *The Gases of the Atmosphere: The History of Their Discovery* (Macmillan, London); 2nd ed. 1902, 3rd 1905, 4th 1915.

Ramsay, W. (1897). An undiscovered gas. *Nature,* **56**, pp. 378–382.

Ramsay, W., Collie, J. N. and Travers, M. (1895). Helium, a constituent of certain minerals. *Journal of the Chemical Society, Transactions*, **67**, pp. 684–701.

Ramsay, W. and Travers, M. (1898a). On a new constituent of atmospheric air. *Proceedings of the Royal Society (London)*, **63**, pp. 405–408.

Ramsay, W. and Travers, M. (1898b). On the companions of argon. *Proceedings of the Royal Society (London)*, **63**, pp. 437–440.

Ramsay, W. and Travers, M. (1898c). On the extraction of the companions of argon and on neon. *Chemical News*, **78**, pp. 154–155.

Ramsay, W. and Travers, M. (1901). On the companions of argon. *Philosophical Transactions of the Royal Society (London) A*, **197**, pp. 47–89.

Rancke-Madsen, E. (1975). The discovery of an element. *Centaurus*, **19**, pp. 299–313.

Remsen, I. (1890). *Inorganic Chemistry*, 2nd ed. (Holt, New York); 5th ed 1907.

Reynolds, J. E. (1895). Argon and the periodic system. *Nature*, **51**, pp. 486–487.

Sedgwick, W. (1895). The existence of an element without valency of the atomic weight of "argon" anticipated before the discovery of "argon" by Lord Rayleigh and Prof. Ramsay. *Chemical News*, **71**, pp. 139–140.

The Times. (1894). The British Association. *The Times* (*London*), 14 August 1894, p. 11.

Thomson, W. (1871). British Association for the Advancement of Science: Inaugural address of the President. *Chemical News*, **24**, pp. 49–56 (at 54).

Tilden, W. A. (1888). *Introduction to the Study of Chemical Philosophy*, 6th ed. (Longmans, Green and Co., London); 11th ed. 1905.

Travers, M. W. (1928). *The Discovery of the Rare Gases* (Edward Arnold, London).

Travers, M. W. (1956). *A Life of Sir William Ramsay, K.C.B., F.R.S.* (Edward Arnold, London).

Trimble, V. (2021). Astronomy meets the periodic table: Or, how much is there of what, and why? In *150 Years of the Periodic Table: A Commemorative Symposium* [Perspectives on the History of Chemistry], eds. Giunta, C. J., Mainz, V. V. and Girolami, G. S. (Springer, Switzerland), pp. 387–407.

Weeks, M. E. (1932). The discovery of the Elements XVIII: The inert gases. *Journal of Chemical Education*, **9**, pp. 2065–2078.

## Select Bibliography

Arabatzis, T. and Gavroglu, K. (2016). From discrepancy to discovery: How argon became an element. In *The Philosophy of Historical Case Studies* [Boston Studies in the Philosophy and History of Science 319], eds. Sauer, T. and Scholl, R. (Springer, Switzerland), pp. 203–222.

Derry, G. N. (1999). *What Science Is and How It Works* (Princeton University Press, Princeton, NJ).

Giunta, C. J. (2001). Argon and the periodic system: The piece that would not fit. *Foundations of Chemistry*, **3**, pp. 105–128.

Kragh, H. (2009). The solar element: A reconsideration of helium's early history. *Annals of Science*, **66**, pp. 157–182.

Kragh, H. (2019). Controversial elements: Priority disputes and the discovery of chemical elements. *Substantia*, **3**(Suppl. 5), pp. 79–90. doi:10.13128/Substantia-740

Kuhn, T. S. (1962). Historical structure of scientific discovery. *Science*, **136**, pp. 760–764.

Labinger, J. A. (2021). The history (and pre-history) of the discovery and chemistry of the noble gases. In *150 Years of the Periodic Table: A Commemorative Symposium* [Perspectives on the History of Chemistry], eds. Giunta, C. J., Mainz, V. V. and Girolami, G. S. (Springer, Switzerland), pp. 303–327.

Lord Rayleigh and Ramsay, W. (1895). Argon, a new constituent of the atmosphere. *Philosophical Transactions of the Royal Society A*, **186**, pp. 187–241.

Oreskes, N. (2019). *Why Trust Science*? (Princeton University Press, Princeton, NJ).

Ramsay, W. (1896). *The Gases of the Atmosphere: The History of Their Discovery* (Macmillan, London); 2$^{nd}$ ed. 1902, 3$^{rd}$ 1905, 4$^{th}$ 1915.

Weeks, M. E. and Leicester, H. M. (1968). *Discovery of the Elements*, 7$^{th}$ ed. (Journal of Chemical Education, Easton, PA).

# Didactic Materials

https://doi.org/10.1142/9789819813858_0011

# INTRODUCTION TO THE DIDACTIC MATERIALS

This section contains materials to help teachers and instructors make use of the episodes from the discovery histories we provide in this book. Indeed, as we stated in the preface of this book, one of the aims of this collective volume is to contribute to making historical scholarship available for teachers and the public and bridge the gap between those who do research and the users of the knowledge that emerged from the research. We believe that dedicating a section of the book to what we call "didactic materials" is one step in this direction.

We gave authors full freedom to choose target audiences, the type of activities they wished to develop for potential students, and the knowledge area they wanted to address. For this reason, we present a variety of materials available for teaching, associated with each of the main chapters in the book. Many authors invite students to reflect and discuss questions related to discovery; others have included clearly defined activities or readings of original literature. Knowledge areas span from the history of chemistry comprehension to understanding the "Nature of Science", to learning chemistry or digging deeper into the field of radioactivity, to reflections on how science and scientists are presented in textbooks and how discovery histories should be taught. Some target pre- or in-service teachers, while others also include activities for their students, mostly at secondary school levels. For the user's convenience, we present target audiences in Table 1 below.

It is recommended that teachers familiarise themselves with the relevant chapters in the book before reading the associated didactic

Table 1. Target audiences for the didactic materials.

| Chapter title | Target audiences |
|---|---|
| Finding Yttrium | Pre-service teachers, secondary school students |
| The Elementary Nature of Chlorine | Pre- and in-service teachers |
| The Many Fathers of Aluminium | Pre-service teachers, secondary school students |
| Discovery of Radium and Polonium in the Era of Radioactivity | Pre-service teachers |
| Clemens Winkler, Cobalt Pigments, and the Discovery of the *Eka*-elements | Pre-service teachers |
| It's Not Elementary — The Case of Jargonium | Pre- and in-service teachers, secondary school students, introductory chemistry students |
| Nobelium, the Questionable Element | Pre-service teachers, secondary school students |
| The Heavy Hydrogen Isotopes | Pre-service teachers, secondary school students |
| Discovering Argon, Helium and Neon | Pre- and in-service teachers, introductory chemistry students |

materials. In some of the chapters in this section, a summary of the connected chapter is presented; however, we believe that these summaries alone do not do justice to the richness of the historical cases. Teachers are also encouraged to explore other recommended resources and adapt the activities or questions presented in the different didactic chapters to their own classroom context.

Since the volume is not a book on chemical education *per se*, none of the tasks have been systematically tried out on target groups and we, as editors, welcome feedback from those who test them, so that we can continue to improve didactic sections of other books in this book series. We also take the opportunity to thank the excellent reviewer who made suggestions that greatly improved these materials.

https://doi.org/10.1142/9789819813858_0012

# FINDING YTTRIUM

Charlotte A. Abney Salomon

## Target Audiences

Pre-service science teachers and secondary school students.

## Background for Pre-Service Science Teachers

A key component of elementary and secondary science instruction, in addition to the specific content knowledge, is the teaching of scientific methods and, more broadly, what state and national standards term the "Nature of Science". However, the narrative and examples that chemistry education often presents to illustrate the history of the field convey a different style of knowledge formation than the method practiced by working scientists, both historically and in the modern laboratory.

In many textbooks, the major discoveries of chemistry are presented as parallel "data points" and appear to have been accepted without question or need for confirmation from the year of a single publication. Students might miss the role of the scientific community when they learn this conventional narrative of the history of chemistry. Very little is shown of peer review processes in action, much less the years of contention and attempts at replication that are often required before the results are widely accepted, or later revisions according to changes in scientific theory or taxonomies.

The "discovery" of yttrium, told in the chapter "Finding Yttrium: Johan Gadolin and the Development of a Discovery", is one example of an ambiguous or complicated historical chemical event that makes clear that simplifying scientific history hides the work of many other researchers, as well as the cooperation and consensus-building required to establish a new scientific fact. The story of yttria and yttrium is a story of the collective classification and reclassification of matter by a group of people connected by letters and journals, with no single clear moment at which the yttrium metal we recognise today was discovered.

The following passage is provided as a classroom activity for secondary school students of introductory chemistry. After reading the text, students can expand upon its ideas and their broader application by developing answers to the questions that follow.

## The Complicated Story of Yttrium

Chemistry textbooks often state that Johan Gadolin discovered yttrium in 1794 when he published a paper documenting a brand new substance that no chemist had written about before, found in a stone from a quarry in the Swedish town of Ytterby. However, the new substance Gadolin described in 1794 was not a metal but an "earth", a type of element that no longer existed 30 years later. When another chemist, Humphry Davy, demonstrated that all earths were actually compounds, each earth became known as the oxide of a (new) metal element instead. In later years, Gadolin's earth, known as yttria, was declared not even a single compound substance but a mixture of the oxides of no less than ten virtually identical metals, or "rare earths". Today's yttrium oxide ($Y_2O_3$) and the metal element yttrium bear little resemblance to Gadolin's yttria.

The story of yttrium also includes a number of other people. Carl Axel Arrhenius selected the stone in the quarry. Anders Gustaf Ekeberg replicated the study, named the mineral *yttria*, and identified other components. Friedrich Wöhler was recognised in 1828 as the first to isolate metallic yttrium from the earth. In 1843, Carl Gustaf Mosander separated Gadolin's yttria, now considered yttrium oxide, into three component oxide substances, naming them yttria, terbia and erbia; other chemists later divided those three into even more compounds. Gadolin did publish the discovery of a new substance in 1794, but many researchers since have also contributed to the modern understanding of yttrium.

## *Discussion Questions for Students of Introductory Chemistry*

1. How would you describe Gadolin's contribution to the discovery of yttrium in your own words?
2. Many other elements have histories that are as complicated as yttrium, or even more so. What are a few ways in which more complicated discovery stories change the common understanding of how facts are established in science? Discuss with your peers.
3. Select and research another chemical element established after 1700, perhaps from another chapter in this book. How many people can you find who contributed to the discovery by predicting, recognising, naming, testing, describing, isolating, or subdividing it? List their names and contributions and present the results to your peers.

https://doi.org/10.1142/9789819813858_0013

# THE ELEMENTARY NATURE OF CHLORINE

Pere Grapí

## Target Audiences

The questions for reflection listed below are intended mainly for pre-service and in-service upper secondary school science teachers.

## Tasks

Read the chapter "The Long and Arduous Pathway Towards the Elementary Nature of Chlorine" before you start working on the tasks.

### *Recall Questions*

1. a) How would you explain what "phlogiston" was to your students?
   b) Why was chlorine first named "dephlogisticated muriatic acid"?
   
   Complete your answer by filling the blank boxes in the diagram below representing Carl Wilhelm Scheele's interpretation of the isolation of chlorine.

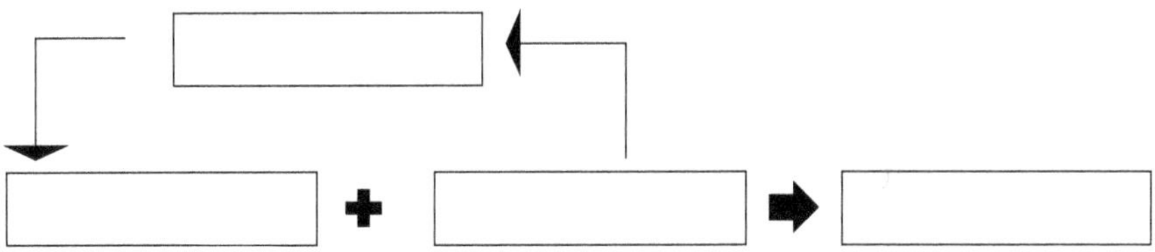

2. Colour was considered, over a long period of time, of primary importance in the naming of chemical substances.
   Who gave the name "chlorine" and why to Scheele's "dephlogisticated muriatic acid"? What name would you give it?
3. Why did Joseph Luis Gay-Lussac and Louis Jacques Thénard refrain in 1809 from public acceptance of the hypothesis of the simple nature of chlorine?
   Think about the benefits and disadvantages of patronage in science and share your opinion with your peers.
4. Why was there so much confusion about whether chlorine was a simple or a compound substance?
   Write down a map with the actors involved in this issue and discuss it with a peer.
5. The perception of chemical similarities between chlorine and iodine was a key point for accepting the elementary nature of chlorine.
   Mention some of the macroscopical, molecular and atomic properties characteristic of chlorine and, on this basis, give an explanation of why chlorine is placed together with other elements in the same group of the periodic table.

## *Explore and Discuss*

6. Joseph Black's experiments on the question of causticity led him to conclude that:
   a. The calx obtained by calcinating an ounce of limestone weighed less than the salt formed by treating an ounce of limestone with vitriolic acid.

b. Fixed air could be transferred from the salt of tartar to limestone.

Interpret both conclusions in modern terms on the basis of information available in Footnote 3 in said chapter about the elementary nature of chlorine.

Recommendation: To familiarise yourself with the old chemical nomenclature — it may be helpful to first rewrite the chemical equations in Footnote 3 using diagrams like the one sketched in Question 1b.

7. Bike tyres lose pressure regularly even when not in use and balloons deflate over time. These are everyday experiences that can help us understand the difficulties in gas handling.

   Look at Figures 2 and 5 in the mentioned chapter. There, you can see that Scheele and Cavendish employed bladders in their experiments. Discuss the pros and cons of using bladders instead of glass flasks for collecting and transferring gases.

https://doi.org/10.1142/9789819813858_0014

# THE MANY FATHERS OF ALUMINIUM

Sarah N. Hijmans

## Introduction and Target Audiences

This didactical material accompanies the chapter "The Many Fathers of Aluminium: Collaboration, Credit and the Construction of Discovery" and contains ideas for a class aimed at secondary school students or pre-service science teachers. It consists of three parts: firstly, the students are asked to provide their own view of scientific discoveries and to briefly research the existing stories of the discoveries of aluminium; the second part consists of a study of the contents of the said chapter and can be adjusted depending on the level of the students; finally, in the third part, the students will be asked to reflect again on the nature of science and scientific discoveries and evaluate what they have learned.

This didactical material might be combined with that of "Finding Yttrium", which similarly provides students with a complex historical case of discovery and asks them to reflect on the retrospective attribution of credit.

## Part I. Preliminary Questions and Exercises for the Students

1. What is a scientific discovery? Can you name any famous discoveries?
2. On the basis of the examples you named, can you identify criteria that are generally used to decide who deserves credit for a scientific discovery?
3. Search for information about aluminium and its history on the Internet or in an encyclopaedia. What can you find about the discovery of aluminium? Make a list of the people who could be considered discoverers of aluminium and why.
4. Study for the opening page Joseph W. Richard's book *Aluminium: Its History, Occurrence, Properties, Metallurgy and Applications, Including its Alloys* (1896), presented in Figure 1.
   a. Who are the people it is dedicated to? Do you agree with the way they are identified?
   b. Look at the phrase at the bottom of the page and think about the uses of aluminium in your everyday life. In what ways may scientific research on aluminium have "benefited mankind"?

## Part II. The Story of Aluminium (Explanation for Teachers)

Depending on the level of the students, they can either study the chapter "The Many Fathers of Aluminium: Collaboration, Credit and the Construction of Discovery" in this book themselves or the teacher might retell the story to them. The timeline below can either be used as a tool to help summarise the story (Figure 2) or as an exercise for the students (Figure 3). In the latter case, students might be asked to fill in the blanks in order to create a chronological overview of the different events that compose the discovery of aluminium.

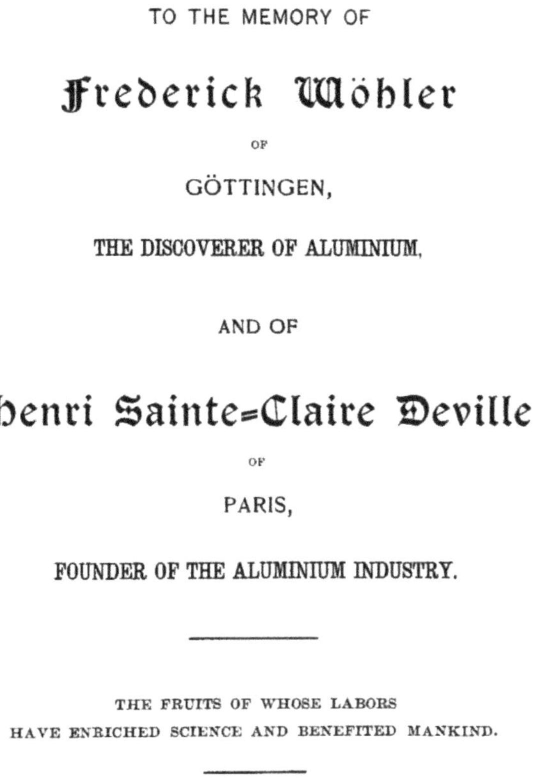
TO THE MEMORY OF

Frederick Wöhler

OF

GÖTTINGEN,

THE DISCOVERER OF ALUMINIUM,

AND OF

Henri Sainte-Claire Deville

OF

PARIS,

FOUNDER OF THE ALUMINIUM INDUSTRY.

THE FRUITS OF WHOSE LABORS
HAVE ENRICHED SCIENCE AND BENEFITED MANKIND.

Figure 1: Dedication of a monograph on aluminium by Joseph Richards (1896). In the public domain.

## Part III. Reflection: Questions for the Students

5. Look at the timeline of the discovery of aluminium (Figures 2 and 3). How does this fit your previous description of scientific discoveries in Questions 1 and 2? Name 2–3 characteristics of scientific practice that you have learned from the story about the many fathers of aluminium.
6. Do you think other element discoveries have a similarly complicated story? Why or why not? Discuss with your peers.

Figure 2: Summary of the events that compose the discovery of aluminium. Engravings credit: Wellcome collection.

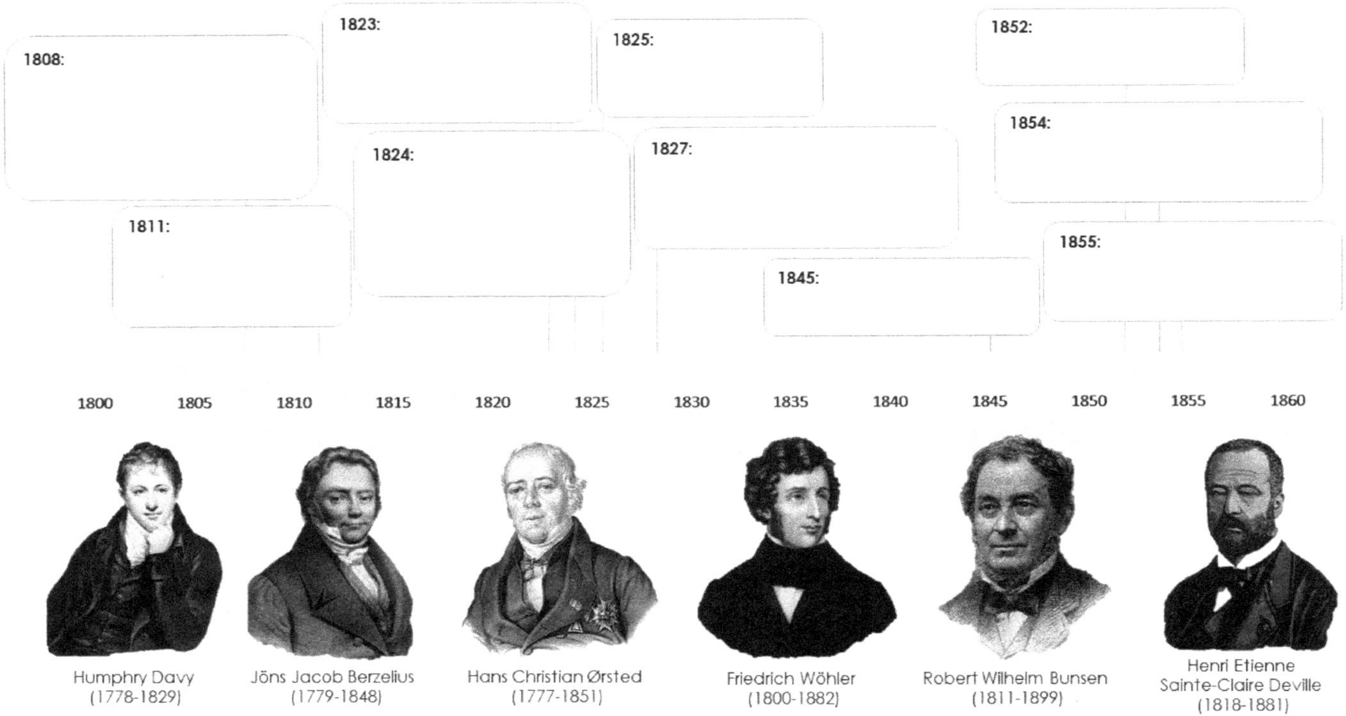


Figure 3: Exercise to create a timeline of the events that compose the discovery of aluminium. Engravings credit: Wellcome collection.

https://doi.org/10.1142/9789819813858_0015

# DISCOVERY OF RADIUM AND POLONIUM IN THE ERA OF RADIOACTIVITY

Annette Lykknes

## Target Audience

Pre-service science teachers.

## Discussion Points

### *Nature of Science*

1. What, in your view, characterises scientific work, based on your reading of the chapter "The Chemistry of the Imponderable? Radium, Polonium and the Discovery of Elements in the Era of Radioactivity"?
2. According to Douglas Allchin, mythical narratives are characterised by heroisation of historical actors, idealisation of scientific processes, the creation of linear timelines of what was originally a web of history, and a sharpening of actors' roles for the sake of telling a good story.

   a) Do you agree that it is problematic that narratives are mythical in Allchin's sense of the word? Why do you think people keep telling such stories if they do not reflect the truth?

b) Look for examples of Allchin's mythical aspects in the chapter on polonium and radium (as well as in other chapters you choose to read). Write an email to your instructor on how to go about engaging students in these topics to avoid disseminating myths about the nature of science.

### *The Discovery Process*

3. Discuss with your peer: When did the discovery of polonium and radium take place, in your opinion? Why?
4. Look up Edmond Rancke-Madsen's criteria for element discoveries: Rancke-Madsen, E. (1976). The discovery of an element, *Centaurus*, **19**, pp. 299–313. Similar rules are also mentioned in the chapter "The Heavy Hydrogen Isotopes: Anticipations and Discoveries" in this book.
   a. At which point in the discovery processes of polonium and radium were these criteria met, and why?
   b. Was your answer to Questions 3 and 4a the same? If no, which criteria did you use for your answer to Question 3? After reading Rancke-Madsen, would you revise your answer to Question 2? To what extent are defined criteria useful to assess past discoveries?
5. Which detection methods were used to identify polonium, radium and other radioelements? Do you have experience with any modern detection method? What are the strengths of the different methods?

### *History of Science in Textbooks*

6.
   a) Look at science textbooks from a grade level of your choice. Which discovery histories can you find? What characterise these narratives?
   b) Based on your reading of *Discovering the Elements*, would you suggest any improvements to the textbook accounts?

Phrase some comments you would want to direct to the authors of these textbooks.

7.
   a) Which (historical) scientists are mentioned in the textbooks you selected? Which countries do they represent? How many women are mentioned by name? You are welcome to comment on this in a modern (political) context.
   b) What kind of message does the presence or non-presence of women in the texts send to you and science students in general?

## *Radioactivity*

8. Take a close look at the periodic table and charts of nuclides. Identify radioelements in the periodic table and look up some of them in the charts of nuclides. How do you think scientists in the early 20$^{th}$ century were able to discover these nuclides/radioelements and their decay series? (Suggested reading: Lykknes, A. and Van Tiggelen, B. (2019). Introduction. In *Women in their Element*, eds. Lykknes, A. and Van Tiggelen, B. (World Scientific, Singapore), pp. 24–36).
9. Read Maria Rentetzi's article about safety measures Marie Curie took for her laboratory: Rentetzi, M. (2017), Marie Curie and the perils in radium, *Physics Today*, 7 November (https://physicstoday.scitation.org/do/10.1063/pt.6.4.20171107a/full/). Try to find out how common radiation poisoning was at her time. Also, look for information about Curie's own illness. Discuss with your peers: Was it risky/dangerous to be a researcher in radioactivity at the beginning of the 20$^{th}$ century?

## *Women in Radioactivity and Chemistry*

10. Radioactivity attracted many women scientists in the early 20$^{th}$ century compared to other fields in science. Explore how that

could be. (Suggested reading: Lykknes, A. and Van Tiggelen, B. (2019). Introduction. In *Women in their Element*, eds. Lykknes, A. and Van Tiggelen, B. (World Scientific, Singapore), pp. 24–36.)

11. Read Annette Lykknes and Brigitte Van Tiggelen's article about the many women in the history of the periodic system (Lykknes, A. and Van Tiggelen, B. (2019). In their element: Women of the periodic table, *Science in School,* 47 (https://www.scienceinschool.org/article/2019/their-element-women-periodic-table/). List the women's names and main contributions. Which of these were you familiar with already?

### Science and War

12. Search the Internet to explore what the Manhattan Project entailed. What was the purpose of the program? Search for the names of women who participated in the project and look for what roles they played. (Suggested reading: Robinson, A. (2019). Isabella L. Karle and the synthesis of plutonium chloride. In *Women in their Element*, eds. Lykknes, A. and Van Tiggelen, B. (World Scientific, Singapore), pp. 374–381.)
13. Investigate which role polonium played in investigations on atomic weapons by reading about its properties and searching the Internet for "polonium" and "war". Can you find other war-related uses of polonium in your search?

https://doi.org/10.1142/9789819813858_0016

# CLEMENS WINKLER, COBALT PIGMENTS, AND THE DISCOVERY OF THE EKA-ELEMENTS

Gisela Boeck

## Introduction and Target Audiences

This material targets pre-service teachers. It consists of three main parts: one about blue colour works, one about Clemens Winkler (storytelling), and one about Dmitrii Ivanovich Mendeleev's 1871 periodic table.

## I. Blue Colour Works

### *Introduction*

Clemens Winkler worked for many years in a blue pigment factory. Blue pigment, called smalte, was produced there. This is the oldest known cobalt pigment. The blue colouring of glasses, faience and porcelain decisively increased their market value. Empirically, alchemists had discovered that some ores could be used for this purpose. Today, we know that these ores contain the element cobalt.

## I.1. *Explore the Discovery Process and the Use of Cobalt (Co) Compounds*

Divide into four groups. Use books and the Internet to solve your task. Prepare to present your results in the plenary.

I.1.1. The first group will look for the meaning of the name cobalt.
I.1.2. The second group researches the discovery of cobalt.
I.1.3. The third group collects illustrations or real glass or porcelain dyed with cobalt.
I.1.4. The fourth group compiles what other chemical compounds played a role in dyeing blue.

Summarise the results in the plenary.

Hints for instructors:
Ores containing cobalt have been known for a long time. It was also known that they could be used for the production of blue pigments. However, it was not clear which elements the pigments contained. In the Middle Ages, miners suspected that these were ores whose silver content had been destroyed by goblins. Miners had long used the name kobold ore (German for goblin ore). In 1735, the Swedish chemist Georg Brandt detected the metal during the processing of cobalt ores, gave it the name cobalt, and later identified it as a new species of metal (close to the modern understanding of "element").[1] Other blue pigments are Egyptian blue and Prussian blue.

## I.2. *Make Your Own Cobalt Detector*

Cobalt can be detected in a test with a phosphorus salt bead with the composition $NaNH_4HPO_4 \cdot 4H_2O$.

[1] We are thankful to Hjalmar Fors for his thoughtful advice on this matter.

*You need:*

- Bunsen burner
- Magnesia stick
- Ammonium sodium phosphate, $NaNH_4HPO_4 \cdot 4H_2O$
- Co-containing salt, like cobalt(II)chloride, $CoCl_2$

*Instruction*

Anneal the tip of the magnesia stick and press it into the phosphorus salt. This causes a small amount of salt to melt, which, when heated, forms a crystal-clear bead. After cooling, the bead is moistened by spraying water on it and bringing it into contact with the finely powdered cobalt salt. It is then heated on the burner. It does not matter whether you work with an oxidising or reducing flame. The blue colouring of the pearls is very quickly visible. See Figure 1.

*Waste management*

The magnesia sticks should be collected in containers for chemical waste.

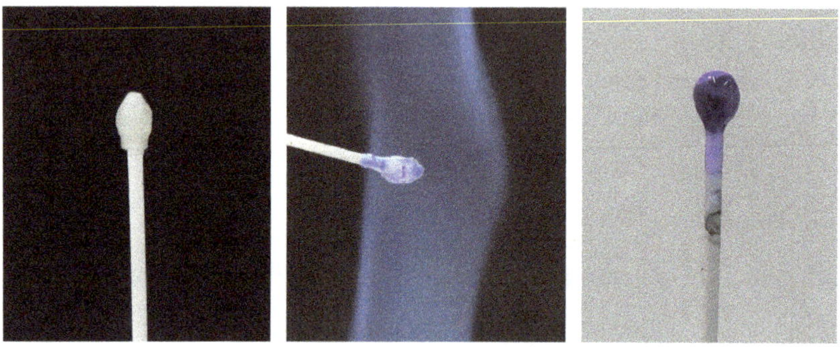

Figure 1: Producing the cobalt detector: Starting with a crystal clear bead, after heating it with $CoCl_2$ in the flame, the blue bead is produced. Courtesy of Edgar Israel Ramirez Tinoco, Institute of Chemistry, University of Rostock.

*Safety instructions*

Cobalt compounds are carcinogenic, especially in powder form, and must be used in very small quantities. They can also cause allergies. The experiment has to be carried out under a fume hood.

Hints for instructors:

When $nNaNH_4HPO_4 \cdot 4H_2O$ is heated, ammonia and water are released. What remains is $(NaPO_3)_x$ with x = 3–8, which produces the characteristic blue colour with cobalt ions:

$$nNa(NH_4)HPO_4 \rightarrow (NaPO_3)_n + nNH_3 + nH_2O$$
$$(NaPO_3)_n + nCoSO_4 \rightarrow nNaCoPO_4 + nSO_3$$

### I.3. *Walk in The Footsteps of Cobalt*

Divide into three groups.

I.3.1. The first group searches the Internet to explore places where blue colour works existed. Mark these regions on a map of Europe. Are some of them located nearby?
I.3.2. The second group should try to find pictures and information about the working conditions in these factories. Try to find out if there are any remains of such factories. Is it possible to visit them?
I.3.3. The third group collects material about the occurrence of cobalt in the world today.

Present your result in the plenary.

Hints for instructors:

I.3.1. Blue colour factories existed for example in Saxony, Thuringia, Prussia, Bohemia, Austria, and Norway

*I.3.2. An example: In the Erzgebirge mountains, there is still a former blue pigment factory. It is located in the small town of Zschorlau. The factory is almost completely preserved. Ultramarine is produced there today.*

*I.3.3. Occurrence today: Congo, Zambia, Canada, Morocco, Cuba, Russia, Australia, and USA.*

## II. Storytelling about Clemens Winkler

### *Introduction*

The teacher presents the history of the discovery of germanium through storytelling. The aim is that this seemingly personal account of Clemens Winkler will help reduce the students' distance to the old and famous chemist.

*Story (Winkler's voice, written by the author)*

Now, I've been at the Bergakademie for 12 years, giving lectures and doing practicals. I have been able to develop further my analytical experience. I was even assigned to examine rock debris from the Krakatau eruption. An iron meteorite lay in front of me. I examined the flue gases. And I did everything myself, every analysis! I heated the furnaces and cleaned everything up at the end. I did not get any assistance, but everything had to be right.

Yesterday, Albin, my cousin, came to see me and brought me some of the new mineral, which is so silvery. Theodor had already shown that it contains a lot of silver. But strangely, he can't get any further with the analysis. Something is wrong. There is always a difference to 100%. Now, he wants me to get to work because he knows that I am very conscientious in my analyses.

So quickly to the lab and to work. But what is it? The analysis values really do not add up to 100%! What have I done, where is my mistake? I need to continue working...

### 4 months later

And again and again, the same result. I am missing 7% in the analysis result. I can't go on, what am I doing? Or is there something completely different at play here — a new element, for example?

But how can I isolate it? Today, I try to mix soda and sulfur with the argyrodite and heat it vigorously. And then I dissolve everything. And now, I add a good shot of hydrochloric acid to the filtrate. What is this? I can't believe my eyes, a sulfide precipitates. But it can't be arsenic or antimony sulfide.

It must actually be a new element. I call it germanium. And now I will sit down at my desk and write an article about this new element right away. I don't know much about its properties yet. In some ways, it is very similar to antimony. Recently, I read something about a periodic table. I haven't quite understood the connections yet. But there was a table by a Mr. Mendeleev with some blanks. These were intended for undiscovered elements. I'll have to take a look at it again. Ah, here my germanium is similar to antimony. It must be the *eka*-antimony.

### 2 months later

There was a lot of excitement after the article about germanium was published. Many people wrote to me, even Mendeleev. Victor von Richter and Lothar Meyer, however, said that the germanium was the *eka*-silicon. Meyer has even published a curve on this, which not only shows the correlation between atomic volume and atomic weight. It is impressive. He even noted properties. And in this curve, I find the right place, yes, it is the *eka*-silicon.

What exciting weeks these have been. But I have to keep working to determine the properties of germanium even more precisely. I wonder what people will use this element for in the future. I did a lot on my own, but I couldn't have done the classification in the periodic table on my own. Like many things, it is a joint effort!

### II.1. *Discuss*

Discuss how you would proceed in a chemical analysis, mentioning a small difference in the result.

### II.2. *Suggest*

Suggest why Winkler named the new element germanium (after part II).

### II.3. *Answer*

Answer Winkler's question about the use of germanium (after part III).

Hints for instructors:

II.2. Freiberg belongs to Germany. Two other new elements were already named after the country or the region where they were found (gallium and scandium).

II.3. Germanium is used in electronics, solar cells, glass, fibres, and nuclear medicine.

## III. Mendeleev's Table from 1871

*Introduction*

The overview of the elements published by Mendeleev in 1871 is very unusual from a contemporary perspective (Figure 2). This table is also called the short periodic system. It has 8 columns and 12 rows. According to Mendeleev, the paired (even) rows contain elements with a metallic character. The unpaired (odd) elements have more acidic properties. Typographically, Mendeleev drew attention to this difference by placing the elements of the even series on the left and those of the odd series on the right. Furthermore, the table shows that Mendeleev had problems with the classification of copper, silver and gold because they appear in the eighth group but also in brackets

**Tabelle II.**

| Reihen | Gruppe I. — $R^2O$ | Gruppe II. — $RO$ | Gruppe III. — $R^2O^3$ | Gruppe IV. $RH^4$ $RO^2$ | Gruppe V. $RH^3$ $R^2O^5$ | Gruppe VI. $RH^2$ $RO^3$ | Gruppe VII. $RH$ $R^2O^7$ | Gruppe VIII. — $RO^4$ |
|---|---|---|---|---|---|---|---|---|
| 1 | H = 1 | | | | | | | |
| 2 | Li = 7 | Be = 9,4 | B = 11 | C = 12 | N = 14 | O = 16 | F = 19 | |
| 3 | Na = 23 | Mg = 24 | Al = 27,3 | Si = 28 | P = 31 | S = 32 | Cl = 35,5 | |
| 4 | K = 39 | Ca = 40 | — = 44 | Ti = 48 | V = 51 | Cr = 52 | Mn = 55 | Fe = 56, Co = 59, Ni = 59, Cu = 63 |
| 5 | (Cu = 63) | Zn = 65 | — = 68 | — = 72 | As = 75 | Se = 78 | Br = 80 | |
| 6 | Rb = 85 | Sr = 87 | ?Yt = 88 | Zr = 90 | Nb = 94 | Mo = 96 | — = 100 | Ru = 104, Rh = 104, Pd = 106, Ag = 108 |
| 7 | (Ag = 108) | Cd = 112 | In = 113 | Sn = 118 | Sb = 122 | Te = 125 | J = 127 | |
| 8 | Cs = 133 | Ba = 137 | ?Di = 138 | ?Ce = 140 | — | — | — | — — — — |
| 9 | (—) | — | — | — | — | — | — | |
| 10 | — | — | ?Er = 178 | ?La = 180 | Ta = 182 | W = 184 | — | Os = 195, Ir = 197, Pt = 198, Au = 199 |
| 11 | (Au = 199) | Hg = 200 | Tl = 204 | Pb = 207 | Bi = 208 | — | — | |
| 12 | — | — | — | Th = 231 | — | U = 240 | — | — — — — |

Figure 2: Mendeleev's table from 1871 [Mendeleev, 1871, p. 151]. In the public domain.

in the first group. Note that didymium turned out not to be an element but a mixture of praseodymium and neodymium.

## III.1. *What Was Possible 150 Years Ago?*

Compare Mendeleev's 1871 table (Figure 2) with a modern periodic table. It contains almost all the main group elements. Check that their assignment to the current main groups I to VII (1, 2, 13–17) is correct. Note that the noble gases were not discovered until the 1890s and group VIII stands for subgroup (not main group) elements.

III.1.1. Transfer Mendeleev's table in a modern form (do not transfer didym!). You may use the blank periodic table provided below (Figure 3, the 7$^{th}$ period is deleted).

III.1.2. Do your and Mendeleev's group numbers agree as regards the main group elements (Is the placement of the main group elements correct?)? Which main group elements were missing (without the 7$^{th}$ period)?

| 1 Ia | 2 IIa | 3 IIIb | 4 VIb | 5 Vb | 6 Vib | 7 VIIb | 8 VIIIb | 9 VIIIb | 10 VIIIb | 11 Ib | 12 IIb | 13 IIIa | 14 IVa | 15 Va | 16 VIa | 17 VIIIa |
|---|---|---|---|---|---|---|---|---|---|---|---|---|---|---|---|---|
| | | | | | | | | | | | | | | | | |
| | | | | | | | | | | | | | | | | |
| | | | | | | | | | | | | | | | | |
| | | | | | | | | | | | | | | | | |
| | | | | | | | | | | | | | | | | |
| | | | | | | | | | | | | | | | | |
| | | | | | | | | | | | | | | | | |
| | | | | | | | | | | | | | | | | |
| | | | | | | | | | | | | | | | | |

Figure 3: An empty periodic table.

| 1 Ia | 2 IIa | 3 IIIb | 4 VIb | 5 Vb | 6 Vib | 7 VIIb | 8 VIIIb | 9 VIIIb | 10 VIIIb | 11 Ib | 12 IIb | 13 IIIa | 14 IVa | 15 Va | 16 VIa | 17 VIIIa |
|---|---|---|---|---|---|---|---|---|---|---|---|---|---|---|---|---|
| H | | | | | | | | | | | | | | | | |
| Li | Be | | | | | | | | | | | B | C | N | O | F |
| Na | Mg | | | | | | | | | | | Al | Si | P | S | Cl |
| K | Ca | | Ti | V | Cr | Mn | Fe | Co | Ni | Cu | Zn | | | As | Se | Br |
| Rb | Sr | Y | Zr | Nb | Mo | | Ru | Rh | Pd | Ag | Cd | In | Sn | Sb | Te | I |
| Cs | Ba | La | | Ta | W | | Os | Ir | Pt | Au | Hg | Tl | Pb | Bi | | |
| | | | | | | | | | | | | | | | | |
| | | | Ce | | | | | | | | | | Er | | | |
| | | | Th | | U | | | | | | | | | | | |

Figure 4: The periodic table in a modern form with the elements known already at Mendeleev's time.

III.1.3. In Mendeleev's table, is it possible to clearly distinguish the elements arranged in the groups on the left or right into main and subgroup elements?

<u>Hints for instructors:</u>

III.1.1. The elements in figure 4 above are in Mendeleev's table from 1871.

III.1.2. The group assignments of the main group elements are already correct. Ga, Ge, Po, and At were, however, missing.

**Tabelle II.**

| Reihen | Gruppe I. — $R^2O$ | Gruppe II. — $RO$ | Gruppe III. — $R^2O^3$ | Gruppe IV. $RH^4$ $RO^2$ | Gruppe V. $RH^3$ $R^2O^5$ | Gruppe VI. $RH^2$ $RO^3$ | Gruppe VII. $RH$ $R^2O^7$ | Gruppe VIII. — $RO^4$ |
|---|---|---|---|---|---|---|---|---|
| 1 | H = 1 | | | | | | | |
| 2 | Li = 7 | Be = 9,4 | B = 11 | C = 12 | N = 14 | O = 16 | F = 19 | |
| 3 | Na = 23 | Mg = 24 | Al = 27,3 | Si = 28 | P = 31 | S = 32 | Cl = 35,5 | |
| 4 | K = 39 | Ca = 40 | — = 44 | Ti = 48 | V = 51 | Cr = 52 | Mn = 55 | Fe = 56, Co = 59, Ni = 59, Cu = 63 |
| 5 | (Cu = 63) | Zn = 65 | — = 68 | — = 72 | As = 75 | Se = 78 | Br = 80 | |
| 6 | Rb = 85 | Sr = 87 | ?Yt = 88 | Zr = 90 | Nb = 94 | Mo = 96 | — = 100 | Rn = 104, Rh = 104, Pd = 106, Ag = 108 |
| 7 | (Ag = 108) | Cd = 112 | In = 113 | Sn = 118 | Sb = 122 | Te = 125 | J = 127 | |
| 8 | Cs = 133 | Ba = 137 | ?Di = 138 | ?Ce = 140 | — | — | — | — — — — |
| 9 | (—) | — | — | — | — | — | — | |
| 10 | — | — | ?Er = 178 | ?La = 180 | Ta = 182 | W = 184 | — | Os = 195, Ir = 197, Pt = 198, Au = 199 |
| 11 | (Au = 199) | Hg = 200 | Tl = 204 | Pb = 207 | Bi = 208 | — | — | |
| 12 | — | — | — | Th = 231 | — | U = 240 | — | — — — — |

Figure 5: Mendeleev's table from 1871 marked with the three predicted elements.

III.1.3. Concerning the division into main group and subgroup elements: Mendeleev was not entirely successful.

## III.2. *Eka and the Right Place*

### *Introduction*

In his work of 1871, Mendeleev described three elements that had not yet been discovered. These are *eka*-boron, *eka*-aluminium and *eka*-silicon.

III.2.1. Check what the word *eka* means.
III.2.2. Mark in Figure 2 the spaces that corresponded to the three *eka*-elements.

Hints for instructors:

Eka is Sanskrit and means "one". Eka-boron is one place under boron.
Figure 5 above shows the three places corresponding to the three mentioned elements. Note that you must look at both the left and right sides of the column.

## III.3. *Atomic Weight as a Tool to Position Elements*

*Introduction*

When Clemens Winkler discovered the new element germanium, he was uncertain about its place in the periodic table. He wrote: "After several weeks of painstaking research, I can now state with certainty that the argyrodite contains a new element, very similar to antimony but very different from it, to which the name germanium may be attached."

III.3.1. According to this description, where in the periodic table would you place germanium?
III.3.2. According to which criteria did Mendeleev and Meyer arrange the elements in the periodic table?
III.3.3. What conclusions do you draw from this for Winkler's further investigations?

Hints for instructors:

III.3.1. By this description, it could also be eka-antimony.
III.3.2. Meyer and Mendeleev arranged the elements in order of increasing atomic weight and properties. Eka-antimony should have an atomic mass of 140 or higher.
III.3.3. Winkler, therefore, had to determine the atomic weight of his new element.

## III.4. *How Accurate Were The Predictions?*

*Introduction*

Knowledge of physical properties needs high-developed instrumentation. Instruments are still being perfected. Nevertheless, there were already astonishingly good results 150 years ago.

III.4.1. Look up the most up-to-date data on germanium in a modern textbook. Complete Table 1.
III.4.2. Discuss the accuracy of Mendeleev's predictions and Winkler's findings in comparison with modern data.

Table 1: Some properties of germanium.

| | Mendeleev's predictions | Winkler's findings | Data in modern sources |
|---|---|---|---|
| Atomic weight/mass | 72 | 72,32 | |
| Density in $g/cm^3$ | 5,5 | 5,469 | |
| Valency | IV | IV | |
| Colour | Dark grey | Greyish white | |
| Melting point in °C | Hard to melt | ≈ 900 | |

Table 2: Some properties of germanium — completed.

| | Mendeleev's predictions | Winkler's findings | Data in modern sources |
|---|---|---|---|
| Atomic weight/mass | 72 | 72,32 | 72,630(8) |
| Density in $g/cm^3$ | 5,5 | 5,469 | 5,323 |
| Valency | IV | IV | II and IV |
| Colour | Dark grey | Greyish white | Greyish white |
| Melting point in °C | Hard to melt | ≈ 900 | 937,4 |

<u>Hints for instructors:</u>

III.4.1. See Table 2 above for data from modern literature. There are still uncertainties to the fourth decimal place of the atomic masses.

III.4.2. It is amazing how accurately Winkler could determine analytical data and Mendeleev could predict data almost 140 years ago.

https://doi.org/10.1142/9789819813858_0017

# IT'S NOT ELEMENTARY — THE CASE OF JARGONIUM

Brigitte Van Tiggelen

## Introduction and Target Audiences

This material aims to invite students to conduct a historical investigation on the so-called "lost element" "jargonium (Jg)", with the help of their teacher, using material accessible online, including original historical documents. We target a range of audiences, from secondary school students to introductory chemistry students at the university level to pre- and in-service teachers. We also suggest, when possible, to consider partnering with a colleague teaching history, as the reading and interpretation of historical material can benefit from and serve as an illustration to the historical method.

To prepare for the lesson, the teacher should familiarise themselves with:

— the concept of "lost elements" (see below),
— the concept of "forerunner" (see below),
— with the chapter "It's Not Elementary: The Successive Lives of *Eka*-manganese aka Element 43", devoted to what may be referred to as the "forerunners" of technetium,
— (if possible) some background on the history of hafnium (Hf). We recommend the following informative and accessible source:

Kragh, H. (1980). Anatomy of a priority conflict: The case of element 72, *Centaurus*, **23**, pp. 275–301, available at: https://doi.org/10.1111/j.1600-0498.1980.tb00235.x

## Background Information for the Teacher

The term "lost element" was coined by Marco Fontani, Mariagrazia Costa and Mary Virginia Orna in *The Lost Elements: The Periodic Table's Shadow Side* (Oxford University Press, 2014), the aim of which is described by the authors as follows:

> Gathering the stories and the documentation of [these] erroneous, spurious, nonexistent "lost elements" into one place is our attempt to faithfully reconstruct the "scene of the crimes," so to speak. [Fontani *et al.*, 2014, p. xxiv]

Indeed,

> Although today's periodic table presently comprises 118 elements, 114 of which bear definitive names,[1] over the many decades of its creation this was not always so. In fact, there are many more elemental "discoveries" later shown to be false than there are entries in the present table. [Fontani *et al.*, 2014, p. xxiv]

Some of these claims raised intense debates and triggered public retractation, while others were simply met by short, perplexed comments and then silence.

In the common language, a "forerunner" designates "something or someone that acts as an early and less advanced model for what will appear in the future, or a warning or sign of what is to follow".[2]

[1] The book on the lost elements was published in 2014, that is, five years before elements Z = 113, 115, 117 and 118 were officially acknowledged by the IUPAC and named nihonium, moscovium, tennessine, and oganesson, respectively.

[2] https://dictionary.cambridge.org/fr/dictionnaire/anglais/forerunner

When it comes to the discovery stories of chemical elements, the term designates substances claimed to be elementary that are now considered to be previous appearances of a well-established element. For instance, "masurium" (claimed to be discovered in 1925) but also "davyum" are now presented as potential forerunners to technetium (officially named and recognised in 1947). Discussing previous claims requires chemists to look back at literature of the past equipped with contemporary knowledge; therefore, forerunners can only be seen as such once the chemical community has reached a consensus on a new element.

In contrast to the label "lost elements", the term "forerunners" indicates that the knowledge, practical work, and ensuing claims of discovery are not completely discarded but reevaluated retrospectively in the light of contemporary knowledge. In some cases, these retrospective assessments even trigger discussions of priority and reflections around on whether the earlier instance should take precedence or not.

## Aims for the Lesson

With the case of Jg, the students should learn to

- search for and use historical material
- think critically about what substantiates a discovery claim in the context of the time
- question the notion of Jg as a forerunner of Hf
- examine the use of history and scientific publications from a remote past by scientists.

## Tasks

### *Initial Exploration*

As a first task, students are asked to find what "jargonium" is. This can be given as a short homework. An Internet quest leads quite straightforwardly to the Wiki page that defines jargonium as a "rejected name for hafnium".[3]

### *Group Discussions*

When the teacher gathers answers from the students next, they might challenge them on the meaning of this short definition. The questions below might be raised. We suggest they be discussed briefly in groups by using the Internet to help them find the answers. The idea is to let the students realise that short answers do not suffice — indeed, more questions arise when one tries to dig into what at first glance may look like a question with a clear-cut answer.

— what is hafnium, and where is this element situated in the periodic system?
— when and by which means was hafnium discovered?
— where does the name hafnium come from, and why was the name chosen?
— which other names were suggested and/or rejected, if any?

Each group can then present their findings briefly to the whole class, and the teacher might then raise the issue and lead a discussion to elucidate how, why and when jargonium was rejected as a name for hafnium. The teacher might take the opportunity to present the notion of a "forerunner" for an element and invite the students to explore the history of jargonium in that context.

## Further Exploration of the Topic and Recommended Resources

Alternatively, the instructor can raise the issue of how, when and why the name jargonium appeared and was later rejected and help students to problematise the question before providing them with

[3] https://en.wiktionary.org/wiki/jargonium

historical material to investigate the question themselves. As a follow-up to the initial discussion, the two texts and further references provided below might be shared with the students to help them investigate why jargonium is considered a forerunner to hafnium. For each document, they can be asked to look for the dates and timeline of the events reported (in particular, in relation to the acceptance of Hf as an element) and which argument is used to make the case that jargonium is a previous occurrence of hafnium. After carefully examining both texts and answering the suggested questions, students should be able to elaborate on how science's past is used.

## 1 – *Forerunners to Hf in "Elementymology & Elements Multidict": A Website Devoted to the Naming and History of Chemical Elements*

This website, created by map and atlas historian Peter van der Krogt between 1999 and 2010, is devoted to the names of chemical elements, looking at both the etymology and the translations in most languages of the world. Each webpage presents a chemical element from the periodic system briefly with an illustration, with some information on the appearance of the element, the minerals in which the element occurs, a portrait of the discoverer(s), and sometimes an example of the use of the element. Since the focus is on the naming, the website also provides a history of the discovery and how the name came about. It furthermore systematically provides lists of "forerunners" for each chemical element embedded in the narrative or simply as bullet points, and the students are invited to analyse the entry for hafnium available at: https://elements.vanderkrogt.net/element.php?sym=Hf

When exploring the webpage, students will find the following text as a beginning to the history and etymology of Hf:

> "Chemists suspected for a long time that zirconium minerals contained the impurity of some unknown element. Many of them reported the discovery of this element, e.g.,

- 1845: The Swedish chemist and mineralogist Lars Fredrik Svanberg (1805–1878) reported the discovery in zircon of this new element, which he named Norium.
- Ostranium (остраний)
- 1869: Henry Clifton Sorby discovered Jargonium by spectroscopical analysis. He, however, retracted his claim the following year, since the spectrum he had seen was due to impure Zirconium.
- 1879: Tellef Dahl, inspector of the Norwegian mines, found a new metal in nickel pyrite of Kragerø, in Norway, and named it Norwegium (Ng) (Norvégium) after the country in which it was first found.
- 1911: "Nigrium"
- "Euxenium" (эвксений)."

(The spelling has been kept as on the website, hence the capitals at the beginning of the elements.)

*Possible questions to students on this text for a more detailed homework*

— Analyse the use of the terms "discovery" or "discover" and other words used to describe previous discoveries of Hf in the excerpt. Note that everything in the excerpt above is written in the indicative, leaving no uncertainty.

— Some more "forerunners" or "lost elements" than jargonium might appear when examining this resource. What can you find about them? For some of them, there is not even a name of the discoverer nor a place or a date, not to mention references or circumstances. What do you think is the value of these items in such a list? How does such a list compare to the canons of scientific or historical writing?

— In the case above, some details are provided. What seems to be the main argument that the "forerunners" listed here can be seen as former appearances of hafnium? And how were these claims dismissed?

## 2 – *Hafnium or Jargonium? A Letter to the Editor Claiming Priority for Jargonium in 1923*

In 1923, the Canadian mineralogist Thomas Leonard Walker authored a letter to the editor of *Nature* more than a year after the announcement that the element Z = 72 had been detected: Walker, T. L. (1923), Hafnium or Jargonium, *Nature*, 112, p. 831, available at: https://www.nature.com/articles/112831e0

Walker makes the claim that Sorby could have had priority. The brief letter is fully reproduced hereafter.

> "Hafnium or Jargonium
>
> The recent discovery of hafnium in minerals containing zirconium serves to remind us of the discovery of jargonium by Sorby in 1869 (*Chem. News*, vol. 20). He found that many zircons contained as much as 10 per cent. of the new element. The two closely related elements, zirconium and jargonium, could be most readily distinguished by spectroscopic methods. Sorby and Forbes found that there was such a marked difference in the solubilities of the chlorides in strong hydrochloric acid, that it was possible to make a qualitative separation. Three years later Cochran investigated this subject and suggested that zirconia and jargonia were identical. My object in bringing this matter before readers ot NATURE is to suggest that the work of Sorby may possibly entitle him to rank as the discoverer of the new element of atomic number 72, and that jargonium may have priority over hafnium and celtium.
>
> T. L. WALKER.
>
> University of Toronto."

*Possible questions to students on this text for the homework*

— Analyse the line of reasoning that leads Walker to conclude that Solby is the true discoverer of element 72. Pay attention to the use of the indicative and the conditional and contrast this presentation with the previous listing of potential former discoveries of that element.

- What is the argument, what is the piece of evidence in the eyes of the author, and to what extent did he use references to scientific published material? For more information, you may access the original publication by Sorby, H.C. (1869), On jargonium, a new elementary substance associated with zirconium, *Proceedings of the Royal Society of London*, **17**(1868–1869), pp. 511–515, available at: https://doi.org/10.1098/rspl.1868.0105
- Look up [Kragh, 1980], cited above. (The teacher may either provide the reference for online reading or chosen excerpts.) Would you consider "celtium" as a forerunner?

More context around the claimed discovery of jargonium and ensuing discussion as well as an analysis of social and political factors that prompted a publication before all results had been confirmed can be found on the blogpost by Hijmans, S. (2020), The short history of (the element) jargonium, available at: https://www.jargonium.com/post/the-short-history-of-the-element-jargonium-post-by-sarah-hijmans-1

https://doi.org/10.1142/9789819813858_0018

# NOBELIUM, THE QUESTIONABLE ELEMENT

Ann E. Robinson

## Introduction and Target Audiences

The chapter "Nobelium, the Questionable Element" in this book examines the controversial discovery and naming of element 102, nobelium. It was an element that was created in a laboratory rather than found in nature. The multiple discovery claims, as well as the hasty naming of element 102, illuminate changes that were occurring in experimental chemistry and physics and their impact on international scientific organisations and on the larger political realm during the second half of the $20^{th}$ century.

The following resources provide general information about the discovery of superheavy elements and the international scientific organisations involved in determining if an element has been discovered and in its naming. Student teachers may wish to choose to focus on a particular element or group of elements when viewing the resources or take a more general overview of the superheavy elements. It is recommended that they read the said chapter about nobelium and access the resources listed below before working on the questions. Some questions might also be suitable for secondary school students (see below).

## Resources and Tasks

*Resources on International Scientific Organisations*

1. The International Union of Pure and Applied Chemistry (IUPAC)
    a. The IUPAC website (http://www.iupac.org/) provides an overview of the current bureaucratic structure of the organisation and the work of its committees, divisions and relationships with other organisations.
    b. The IUPAC celebrated its $100^{th}$ anniversary in 2019. A special issue of its journal, *Chemical International*, was devoted to the history of the IUPAC (volume 41, number 3, July 2019). The articles in the issue can be accessed for free on the publisher's website (https://www.degruyterbrill.com/journal/key/ci/41/3/html).
2. The International Union of Pure and Applied Physics (IUPAP)
    a. The IUPAP website (https://www.iupap.org/) provides an overview of the current bureaucratic structure of the organisation and the work of its committees, including its work with other organisations, and some historical information.
    b. A history of the IUPAP from 1923 to 1972 can be found in *Physics 50 Years Later* (The National Academies Press, Washington, DC), which is freely available online (https://doi.org/10.17226/20232).

*Internet Resources to View*

The following videos provide information about how superheavy elements are discovered and provide viewpoints of scientists from the different laboratories involved. Each has an emphasis on a specific element or group of elements.

1. In 1994, the Lawrence Berkeley Laboratory (LBL) produced a roughly 25-minute documentary, "Search for Heavy Elements", that chronicles the discovery at LBL of elements heavier than uranium. The documentary features archival film and

photographs, as well as interviews with scientists, and is suitable for all ages. It contains a very good explanation of how elements are synthesised, with an emphasis on element 106. The documentary can be found on YouTube: https://www.youtube.com/watch?v=xu1qQEZSEaI&ab_channel=BerkeleyLab

2. The Periodic Table of Videos is a project between video journalist Brady Haran and the chemists at the University of Nottingham, most notably Sir Martyn Poliakoff. Most videos are between 5 and 10 minutes long and are intended for a general audience. No special chemical or physical knowledge is expected. There are videos on each element as well as on many periodic table and element-related subjects. There are several videos of particular interest to nobelium and the synthesis of superheavy elements.

    a. "Nobelium" reveals just how little is known about element 102, even half a century after its discovery: http://www.periodicvideos.com/videos/102.htm
    b. "Super Heavy Elements in Darmstadt" is a road trip to GSI in Darmstadt, Germany, where elements 107–112 were discovered and provides an excellent demonstration of how superheavy elements are synthesised: http://www.periodicvideos.com/videos/feature_darmstadt.htm
    c. "The Element Creator" is a tour of the accelerator in Dubna led by Yurii (Tsolakovich) Oganessian: http://www.periodicvideos.com/videos/oganessian_tour.htm
    d. "The Office of Georgy Flyorov" provides an intimate look at both Georgii Nikolaevich Flerov and Oganessian: http://www.periodicvideos.com/videos/flyorov_office.htm

## *Questions for Discussion*

The search for new superheavy elements raises some questions about which would be suitable for both secondary and tertiary students. They can be applied to a particular superheavy element, a set of these elements, or the superheavy elements in general. There are no right answers to these questions, but they should encourage thoughtful discussion. The resources listed above provide additional information not found in the chapter on nobelium in this book that may add depth to the discussion.

1. Why would governments be willing to spend millions of dollars in the attempt to synthesise new chemical elements that have no practical purpose?
2. Historically, the IUPAC has been responsible for officially naming new elements. Given that new chemical elements are now more often synthesised by physicists and through physical methods, should the IUPAC continue to be responsible for the naming of these elements, or should the IUPAP assume the responsibility?
3. After World War II, new chemical elements began to predominately be named after scientists rather than properties of the element, minerals, places, mythological characters, or astronomical objects. Why might this have been the trend? Should chemical elements be named after people? If so, should they be named after living people? Should scientists be the only people considered for new element names?

https://doi.org/10.1142/9789819813858_0019

# THE HEAVY HYDROGEN ISOTOPES

Helge Kragh

## Target Audiences

Students in secondary school or teacher education programs.

## Introduction

The case of the discovery of the two heavy hydrogen isotopes deuterium and tritium, told in the chapter "The Heavy Hydrogen Isotopes: Anticipations and Discoveries", is of interest not only for intrinsic reasons but also because it illustrates more general features in the discovery process, such as the relation between the concepts of anticipation, prediction and discovery. In a nutshell, despite a few speculations still in the late 1920s, it was believed that hydrogen is a simple element with no heavy isotopes. Then, a systematic search led Harold C. Urey and coworkers to conclude in 1932 that they had discovered the $^2H$ isotope (deuterium). The heavier $^3H$ isotope had been hypothesised early on, but it took until 1934 before Ernst Rutherford and his group reported evidence for $^3H$ (tritium) produced artificially in nuclear reactions. As is the case with several element discoveries, the discoveries of the hydrogen isotopes involved physical methods as much as they relied on traditional chemical methods.

## Ideas for Classroom Discussions

1. Many of the 118 elements are known to exist in more than one isotopic form. Why should the discovery of a new isotope be of any interest and perhaps of an interest comparable to that of a new element?
2. Apart from the hydrogen isotopes, there are a few other isotopes of historical significance. Students may be asked which they are and for what reasons they have historical significance. The uranium isotope $^{235}U$ is an obvious candidate. Why? Was it ever discovered in the same sense that deuterium was discovered? The students should be encouraged to explore the question using appropriate sources on the Internet. For a beginning, see the Wikipedia article "Uranium-235".
3. The consequences of a scientific discovery are generally unpredictable at the time of the discovery. No one believed in the late 1930s that tritium was more than a scientific curiosity and yet it was crucial to the construction of the hydrogen bomb and is today equally crucial in the plans to construct fusion reactors as sustainable energy sources for the future. Likewise, deuterium (heavy water) was crucial in the development of the uranium bomb. Students may discuss the connections, for example, by imagining scenarios of counterfactual history: how would modern history have changed had deuterium and tritium not been discovered? Or, more concretely, by choosing an isotope: How would the world today be different if this isotope were not discovered, in your opinion?
4. The strange $H_3$ molecule was part of the story leading to the recognition of tritium. What is the present status of this molecule and its cation $H_3^+$? Again, explore the question by means of the Internet, starting with the Wikipedia articles on triatomic hydrogen and trihydrogen cation.
5. Today, Urey is probably better known for the so-called Urey-Miller experiment than for his discovery of deuterium. What was this experiment about, and why is it famous? Is the name "Urey-Miller experiment" an appropriate eponym?

6. The work of Fred Allison briefly mentioned in the chapter about hydrogen isotopes in this book is today widely considered "pathological science", a term introduced in 1953 by Irwin Langmuir, a chemistry Nobel laureate of 1932. Look up what this is about and discuss whether Langmuir's criteria are acceptable or not. How can "good science" be distinguished from "bad science", alias pathological science?
7. The scale of atomic weight, e.g., that H ≡ 1 or O ≡ 16, is conventional and has changed over time. Why did chemists and physicists for a time adopt different scales? What are the scale and the corresponding unit today, and for what reasons?
8. Gilbert N. Lewis was nominated 14 times for a Nobel Prize in chemistry, mostly for his work in other areas than deuterium and heavy water. He never received the prize. Who was Lewis, and why is he considered one of the most important chemists in the first half of the 20$^{th}$ century?
9. Luis Alvarez and Robert Cornog discovered in 1939 that tritium is beta-radioactive, describing its decay as ${}^{3}_{1}H \rightarrow {}^{3}_{2}H + e^{-}$. Something is missing in the scheme. What?
10. Look up how many times a Nobel Prize has been awarded to the discovery of a new element and/or isotope. Which was the first time and which was the last? Do the data tell us anything about how chemistry (or science generally) has evolved from 1901 to the present?
11. Ernest Rutherford is usually credited for the first detection of tritium nuclei, whereas those who performed the experiment (Mark Oliphant, Paul Harteck) are rarely mentioned. Is this an example of what sociologists of science call the "Matthew effect"? Find out what this term covers and suggest examples of the Matthew effect related to the discoveries of elements.

https://doi.org/10.1142/9789819813858_0020

# DISCOVERING ARGON, HELIUM AND NEON

Carmen J. Giunta

## Summary of the Chapter on Argon, Helium and Neon

Several models of scientific discovery and criteria for discoveries of elements are presented in the chapter "In Defence of a Journalistic Model of Describing Discovery: The Case of Argon, Helium and Neon". The discoveries of three elements — argon, helium and neon — are described and discussed in light of the models and criteria. It is suggested that discoveries of elements are susceptible to description by a journalistic model, that is, by answering the common journalistic questions of who, what, when, where, how, and why. Some of these simple questions have also been associated with the rightly criticised individualist model of scientific discovery. It is argued that the deficiency of the individualist model lies in the assumption or demand that the questions have simple answers.

## Introduction and Target Audiences

Questions 1 and 2 below involve reading beyond the said chapter, and each requires an essay-length answer. I recommend these questions to students of science teaching and to science teachers for the purpose of constructing another "case" beyond the one described in this chapter and for seeing for themselves whether the model is more widely applicable than to argon, helium and neon. These questions, especially the first, could well be used as assignments about the nature of science in a science teaching class. Questions 3 and 4 are also appropriate for university students.

## Questions and Exercises for Further Study

1) Read a substantial account of the discovery of an element, and based on the information presented in the account, answer the six journalistic questions about its discovery: what was discovered, who discovered it, and when, where, how, and why was it discovered? By a substantial account, we mean a reasonably detailed description such as another chapter in this book or an account from Weeks and Leicester [1968] (or an earlier edition of *Discovery of the Elements*),[1] not a "bite-sized" paragraph. Note the questions, if any, for which the chapter contains insufficient information. Do the unanswered questions, if any, seem to be ones the author simply chose not to address, or are the questions difficult or impossible to answer? (You may use additional sources.)
2) The following analysis is about how the accounts are influenced by motivating questions; that is, how stories about ostensibly the same events can differ depending on what question(s) the author(s) set out to answer. Compare an account of the discovery of an element taken from this book with the account of the discovery of the same element in [Weeks and Leicester, 1968] (or an earlier edition of *Discovery of the Elements*). What question(s) did the author(s) set out to answer in each account? That is, what question(s) constituted the main point of each account?

[1] Full text available at: https://archive.org/details/discoveryofeleme0000mary

The guiding question(s) may be one or two of the journalistic questions or may be something else entirely (for example, what technologies enabled the discovery or what societal concerns made the research possible?).

3) In what sense(s) is the discovery of an element never complete? Select an element whose discovery is treated in this book or in another source on element discovery. List two or three physical or chemical properties of the element that were determined at the time of the discovery or shortly thereafter and two or three that were not determined until considerably later or are not yet known.
4) This question broadens the scope of critical reading beyond the discovery of elements to other scientific writings and even to other forms of expository non-fiction writing. Read an article assigned by your instructor. Which of the journalistic questions does the article answer about the matters described in the article? Which does it not answer? Why does it answer some and not others? What is the author attempting to communicate? Are the journalistic questions even relevant to the author's purpose? (Note to instructors: Items 17–19 in *Elements and Atoms* (see Supplementary Materials below) are very brief communications by scientists on aspects of radioactivity from the late 19th and early 20th centuries annotated for the benefit of students. They would be suitable for this exercise even for students not concentrating in science.)

## Supplementary Materials

A public lecture about the discovery of argon delivered by Lord Rayleigh in 1895, extensively annotated by Carmen Giunta to provide background information and bring out the scientific logic of the story, is available online at: https://web.lemoyne.edu/~giunta/EA/RAYLEIGHann.HTML. William Ramsay's speech "An Undiscovered Gas" from 1897 is also available, similarly annotated, at: https://web.lemoyne.edu/~giunta/EA/RAMSAYann.html. These two selections, slightly revised in 2021, are two chapters from a collection of papers on elements and atoms: *Elements and Atoms: Case Studies in the Development of Chemistry* (2002) (editor: Carmen J. Giunta) (https://web.lemoyne.edu/~giunta/EA/index.html, last accessed 28 May 2025).

https://doi.org/10.1142/9789819813858_bmatter

# INDEX

www.ingramcontent.com/pod-product-compliance
Lightning Source LLC
Chambersburg PA
CBHW071113290126
38591CB00018B/43
*9789819813841*